Preface

There are several books on Electrochemistry and most of them include a great deal of introductory theory, which we omitted from our larger book because of space considerations. The net result is that, these books contain the real content of Electrochemistry.

Our purpose in this book, is to meet the needs of teachers who present this subject to students who do not have the time or perhaps the inclination to pursue it in depth, but who may also require explicit coverage of basic topics.

This book provides an authoritative account of every aspects of Electrochemistry of current interest and demonstrates progress in this subject that has been made in the recent past.

Editor

Contents

	Preface	v
1.	Metal and the Metal-Solvent Coupling	1
2.	Electronic Properties of the Metal Surface	26
3.	Metal-Adsorbate Interaction	53
4.	Electrocatalytic Oxidation of Oxygenated Aliphatic Organic Compounds	105
5.	Study of the Adsorbed Species	136
6.	Characterization of the Electrode Material	206

1

Metal and the Metal–Solvent Coupling

A specific role of the metal in many properties of the metal/solution interface has been known for a very long time. Insofar as the capacitance of the ideally polarized electrode is concerned, a possible contribution of the metal was suspected very early. However, the early attempts to include this idea in the modeling of the interface by using the theoretical methods available at that time were unsuccessful. For about the next 50 years, the so-called "traditional approach" was based on a macroscopic picture of the electrode surface. All the models for the capacitance proposed in this approach were based on two premises. The first is that the metal behaves like an ideal conductor. This means, for example, that a possible contribution to the capacitance related to a change of the surface potential of the metal with the applied potential is negligible. The second is that the electrode surface forms a rigid boundary on the solution side. This means that the distance of closest approach of solvent molecules or ions to the "electrode plane" is mainly determined by steric effects related to their size.

In such a macroscopic picture of the electrode, the influence of the nature of the metal could be accounted for only by invoking its influence on the polarization of the solvent via some "residual interaction." The influence of the structure of the electrode surface, for instance, the difference between liquid or solid electrodes or the effect of the crystallographic orientation of the surface, was discussed only in terms of steric effects. Thus, no coherent picture of the role of the metal emerged from this approach.

The impetus for an alternative analysis of the role of the metal did not come from new experimental facts which contradicted the traditional view but rather from the progress that was made in the understanding of the surface properties of metals on a microscopic level in the early seventies. These studies provide information precisely on those aspects for which the classical description of the electrode surface was insufficient. Their first important consequence is that a direct contribution of the metal to the capacitance cannot be neglected. In addition, this description of the metal surface on a microscopic level provides now a new insight into its interaction with the adjacent liquid phase and hence its indirect contributions. These ideas have now received sufficient support from theoreticians. Thus, it seems worth discussing their consequences in the models proposed in this past decade.

A comprehensive discussion of the properties of the metal/solution interface in the light of these developments requires that both sides of the interface be considered. Such a task is beyond the scope of this chapter. The main purpose of this work is to discuss the alternative analysis of the role of the metal which followed from the improved description of the surface properties of metals. Thus, important aspects of interfacial properties will not be covered here. A first limitation is that only capacitance curves will be discussed. Since both their shape and their magnitude depend on the nature of the electrode, they provide a good test for a new analysis of the metal at the interface. Insofar as the solution is concerned, its description in the classical models has been reviewed exhaustively in this series. 1~2 These models will be mentioned only in relation to their description of the metal. Recent progress"' in the statistical mechanical description of the solution is also beyond the scope of this chapter. Other limitations of' our discussion concern the kind of interfaces considered here. Interfaces showing a pronounced specificity related to the nature of the electrolyte or that of the solvent are naturally of the greatest interest. However, at the present stage in the process of reassessing the role of the metal, it is natural to test the theoretical models first on simpler situations such as, for example, interfaces without specific ionic adsorption. Quite generally, only interfaces where the interaction of the metal with the solvent and the electrolyte is weak will be discussed. We will leave out the most interesting case of specific ionic adsorption. Another important point is to select the metals to be discussed. In fact, some recent models have tried to describe the capacitance for a variety of metals, neglecting important questions such as are the capacitance measurements reliable for these metals, and are there sufficient theoretical studies and experimental data on their properties in vacuum? The discussion here will be restricted to cases which meet these requirements. Details of the models for the metal at the interface can be found in the chapter by Goodisman in an earlier volume of this series. Since the latter chapter was written, the role of metal–solvent coupling has been emphasized in several papers. The present work, which, is complementary to the chapter by Goodisman, will be mainly devoted to this aspect and its consequences for the analysis of the charge–capacitance curves of various metals.

Before proceeding further, the well–known difficulty one faces in the interpretation of interfacial properties should be recalled. it is common sense to state that most observed interfacial properties reflect contributions from the metal and the solution interacting with each other. From experimental data, one tries to obtain information on the interfacial structure and the interactions on a microscopic level. Information gleaned in this way concerns the interface as a whole. Therefore, the extent to which more specific information can be extracted from experiment depends on an ad hoc hypothesis on all possible contributions. This is well illustrated by the example of the potential of zero charge E, usually written as:

$$E_{\sigma=0} = \Phi + F\,[\delta\chi^m - g^s_{dip}]_{\sigma=0} - E_T \qquad (1)$$

This relation expresses the potential of zero charge in terms of the electronic work function Φ, the variation $\delta\chi^m$ of the electrostatic barrier at the metal surface χ^m induced by the solution, and the potential drop g^s_{dip} due to the orientation of the solvent dipoles. In this expression, E_T(ref) depends only on the nature of the reference electrode. From this expression for $E_{\sigma=0}$, it is possible to extract information only on the quantity in brackets, Because of the term E_T, the information obtained in this way is only relative, generally referenced to Hg. Since a plot of $E_{\sigma=0}$ versus Φ gives a straight line with a slope less than unity for most *sp* metals, the dipolar term in brackets is metal–dependent. Thus, no definite conclusion regarding g^s_{dip} alone can be obtained from $E_{\sigma=0}$. It follows that different estimates of the terms in $E_{\sigma=0}$ lead to different conclusions about the metal–solvent interaction (see the debate on the "hydrophilic" or "hydrophobic" character of silver in and references therein).

In their turn, charge–capacitance curves relate macroscopic quantities. In order to extract information on a microscopic level from these curves, a model is always required. The link between the macroscopic and microscopic behavior being indirect, it is not possible to

Metal and the Metal–Solvent Coupling

draw definitive conclusions on a microscopic level only from charge–capacitance curves.

In the analysis of the capacitance, the emphasis was placed for many years on the solution side, since it was assumed that the rearrangement of the electronic structure of the metal with the charge has no appreciable effect on the capacitance. Now, this assumption is challenged. In addition to its long known influence, in particular oil the molecular polarization, the metal makes a non–negligible direct contribution to the capacitance. This raises the following question: To what extent will these improvements in Our understanding of the properties of the metal lead to a different interpretation of' experimental observations? In other words, will the separation of metal effects reveal qualitatively new aspects about tile remaining parts of' the interface?

This chapter is organized as follows. In Section II, some experimental data on the capacitance will be reviewed briefly. A formal statement of the problem together with a brief survey of the role of' the metal in the traditional approach will then serve as an introduction to the more recent work. In Section III, some theoretical methods used for the study of metal surfaces will be presented. Some aspects of the theory which may be of interest in areas other than capacitance studies will also be pointed out. In Section IV a brief discussion of the results of experiments on adsorption from the gas phase and some theoretical studies concerning the interaction of adsorbed species with the surface will be presented. In the following section, the general structure of recent models of the capacitance will be discussed. Some models with fixed metal–solvent separation will be detailed in Section VI. A more general situation will be considered in Section VII. The last section will be devoted to the results of a semiempirical analysis of experimental data.

GENERAL ASPECTS

Some Experimental Facts

In the first part of this section, some classical data for the ideally polarized electrode have been selected as an illustration of the influence of the nature of the metal. In the original works, these data were analyzed in the framework of the traditional description of the electrode.

To begin with, it is useful to recall that the study of the electrical properties of the ideally polarized electrode started by considering the electrocapillary curve relating the interfacial tension γ to the electrode potential E. The interfacial tension itself is known to differ little at the mercury/water interface from the surface tension of mercury. Moreover, the variation of y along the electrocapillary curve does not exceed 25% of its value at the potential of zero charge (pzc). Thus, the magnitude of the interfacial tension at the metal/solution interface is largely (75%) determined by the metal contribution. Accordingly, one may think that even a small variation of the metal contribution can be responsible for a large part of the observed variation. This situation holds also for gallium. The electrocapillary curve and the capacitance are related through the Lippmann equation. This link, which provides a possible route for studying the capacitance, suggests that the role of the metal should be considered carefully. This route has not been actually followed, mostly because the calculation of the second derivative of y with respect to potential is difficult.

An important point to mention here is that the widely accepted analysis of the capacitance for nonadsorbing ions proposed by Grahame will be followed throughout this chapter. In this analysis, the measured capacitance is separated into a contribution given by the Gouy–Chapman model and a co concentration–independent part $C_i(\sigma)$, as represented, for example, by Parsons–Zobel plots. Therefore, in this chapter only $C_i(\sigma)$ curves will be discussed. $Ci(\sigma)$ is understood as the high–concentration limit of the measured capacitance at which the Gouy–Chapman contribution vanishes. The interpretation of this separation in terms of "diffuse layer" and "inner layer" contributions has been discussed in some recent work based on statistical mechanics.

In the case of solid electrodes, an important aspect is that the $C_i(\sigma)$ curves are usually deduced from experiment by correcting the measured capacitance by a "roughness factor" R, since the real area of the electrode surface differs from the geometrical one. From capacitance data, the roughness factor can be determined either from Parsons–Zobel plots or from the criterion proposed by Hamelin and Valette that the capacitance should be monotonic near the pzc. The two methods may give slightly different estimates of R, so that the magnitude of the $C_i(\sigma)$ curves deduced from experiment may depend on the method employed to determine R. However, this variation of R does not dramatically affect their magnitude in the cases discussed below. Moreover, the shape of the $C_i(\sigma)$ curve is the same irrespective of which criterion is used to determine R. For a discussion of these points, see, for example, Therefore, although an independent estimation of the roughness factor might be necessary for a quantitative comparison between theory and experiment, this is not strictly necessary for an analysis on a qualitative level, especially that which will be presented in the last section of this chapter.

In Fig. 1, $C_i(\sigma)$ curves obtained with four different metals are shown. They include results from the classical work on liquid electrodes by Grahame (Hg/NaF) and by Frumkin et al. (Ga/Na$_2$SO$_4$) and more recent work on monocrystalline solid electrodes by Hamelin et al. (Au/

NaBF$_4$) and Valette and Hamelin (Ag/ NaF). All these curves are for nonadsorbing electrolytes (note however that there is slight adsorption in the case of Ag/NaF, and in the case of Ga/Na$_2$SO$_4$ the curve at 0.5 N is not corrected for the diffuse layer contribution). Data on Ag/KPF$_6^{22}$ will also be discussed. The curves for solid electrodes are based on the real area. For a review on solid electrodes.

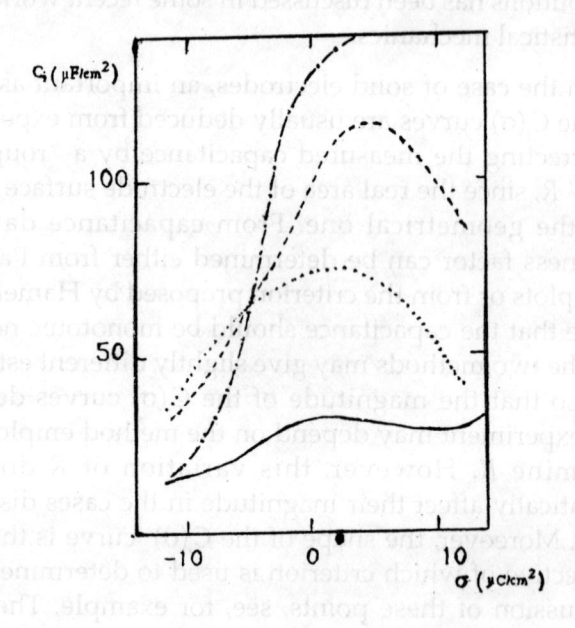

Figure 1. C_i(s) curves for HG, Ga, Ag, and Au: ———, Hg/NaF[19]; ---.---, Ga/Na$_2$SO$_4$; ---, Ag/NaF[21]; . . ., Aug/NaBF$_4$.

The curves in Fig. 1 exhibit appreciable variations both in shape (asymmetric, bell–shaped, or humped) and in magnitude with the nature of the metal. This last point is well illustrated by comparing the values of the capacitance at the pzc. All curves show either a maximum or a hump at a slightly positive charge. Figure I provides then a good illustration of the influence of the nature of the metal. This picture can now be completed by considering two other types of experiments.

The first one concerns the effect of temperature on $C_i(\sigma-)$ curves. In Fig. 2, the classical results of Grahame for Hg/NaF[24] are shown. It is almost -superfluous to recall the amount of work devoted to the interpretation of these curves and to the rather controversial origin of the hump in the low-temperature curves (see, for example, the reviews by Reeves, Habib, and Trasatti in this series). We also mention very recent data on gold and silver obtained by. Hamelin and co-workers.

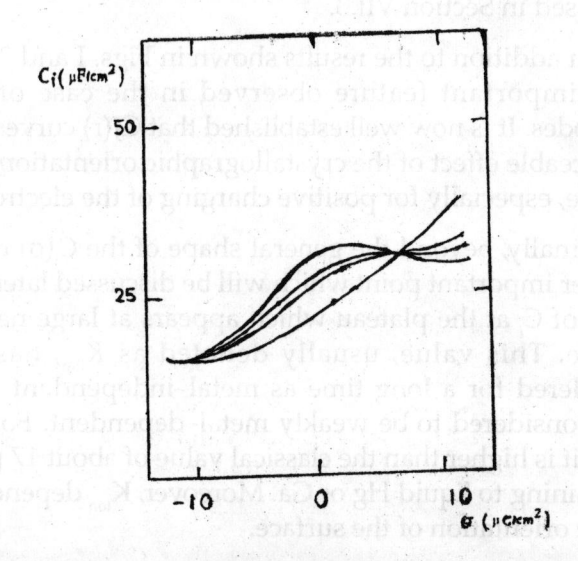

Figure 2. $C_i(s)$ curves at different temperatures for Hg/NaF. From top to bottom at the pzc: $T = 0, 25, 45$, and 85^0C.

Among the effects related to temperature, important results concern the entropy of formation of the inner layer, Sin (see, for example, Fernando Silva). In the case of the Hg/NaF interface, a well-known feature of the $S_{in}(\sigma)$ curve is that it exhibits a maximum at a negative charge. In the models attributing the capacitance hump to dipoles, it was difficult to reconcile in a simple way the occurrence of the hump at a positive charge with the maximum of $S_{in}(\sigma)$ being at a negative charge. However, a dipolar origin of the hump was challenged in the work of Bockris and co-workers by invoking specific ionic adsorption. This was a long debated subject, and we refer the reader to Refs. 1, 2, and 6, cited above. Since none of the related interpretation considers a direct contribution of the metal, this raises the following question: What will happen if the effect of the metal can be separated from capacitance curves? This aspect will be discussed in Section VII.3.

In addition to the results shown in Figs. I and ?, there is an important feature observed in the case of solid electrodes. It is now well established that CJ(r) curves show a noticeable effect of the crystallographic orientation of the surface, especially for positive charging of the electrode.

Finally, beyond the general shape of the $C_i(\sigma)$ curves, another important point which will be discussed later is the value of C_i at the plateau which appears at large negative charge. This value, usually denoted as K_{ion}, has been considered for a long time as metal–independent. K_{ion} is now considered to be weakly metal–dependent. For solid silver it is higher than the classical value of about 17 μF cm^{-2} pertaining to liquid Hg or Ga. Moreover, K_{ion} depends also on the orientation of the surface.

From these data, it appears that the capacitance depends not only on the nature of the metal, but also on the crystallographic structure of its surface. A nontrivial feature of the $C_i(\sigma)$ curves is their strong dependence on temperature. It should be recalled that most of the data presented above have already been discussed in the classical models. However, because of the developments which have appeared in the past decade, these data should be reconsidered in the light of the new description of the metal. Thus, it is useful to discuss both classical models and more recent ones in a common framework if possible, at least from a formal point of view. This will allow us to point out why a more detailed description of the metal than the traditional one is really required.

Formal Expression of the Capacitance

For a planar interface, let $\Delta\phi_s^m$ be the potential drop between two planes, the first one located well inside the metal (at $z = -\infty$) and the second one in the bulk phase of the solution (at $z = +\infty$). Integrating Poisson's equation leads to

$$\Delta\phi_s^m = \phi(-\infty) - \phi(+\infty)$$

$$= -4\pi \left[\int_{-\infty}^{+\infty} z\rho_m(z)dz + \int_{-\infty}^{+\infty} z\rho_s(z)dz + \int_{-\infty}^{+\infty} P(z)dz \right] \qquad (2)$$

In this expression, $P(z)$ is the polarization due to the solvent molecules, in the direction z normal to the interface. It arises both from the permanent dipoles (orientational contribution) and from any induced dipoles (distortional contribution). The term $\rho_m(z)$ is the local density of charge at a position z in the interface, due to the spatial distribution of ionic cores and conduction electrons of the metal, averaged in the x–y plane. The net charge per unit area on the electrode is then $\rho = \int_{-\infty}^{+\infty} \rho_m(z)dz$. The term $\rho_s(z)$ is the corresponding quantity due to the spatial distribution of the anions and the cations of the solution, so that $-\sigma = \int_{-\infty}^{+\infty} \rho_s(z)dz$. Note that the convention for the potential drop employed here is that adopted in classical textbooks of electrochemistry, while in recent models, the potential drop

$\Delta\phi = -\Delta\phi_s^m$ is more often considered. The condition that the interface taken as a whole is electrically neutral implies that $\Delta\phi_s^m$ as defined in Eq. (2) is independent of the origin, of the coordinates on the z axis. The normalized first moments Z_m of $\pi_m(z)$ and Z_s of $\rho_s(z)$ with respect to the same (and yet arbitrary) origin $z = 0$ are defined as

$$Z_m = 1/\sigma \int_{-\infty}^{+\infty} z\rho_m(z)dz \tag{3a}$$

$$Z_m = -1/\sigma \int_{-\infty}^{+\infty} z\rho_s(z)dz \tag{3b}$$

These quantities are often referred to as the "center of mass" or "centroid" of the charge distributions. By using the definitions of Z_m and Z_s in Eqs. (3) $\Delta\phi_s^m$ can be written as

$$\Delta\phi_s^m = 4\pi s(Z_s - Z_m) - 4\pi \int_{-\infty}^{+\infty} P(z)dz \tag{4}$$

and the different capacitance can be written as

$$1/C = \partial\Delta\phi_s^m/\partial\sigma = 4\pi\partial/\partial\sigma[\sigma(Z_s-Z_m)]$$

$$- 4\pi\partial/\partial\sigma \int_{-\infty}^{+\infty} P(z)dz$$

Let us define some other useful quantities. We first introduce a quantity $X^m(\sigma)$ defined as

$$\chi^m(\sigma) = -4\pi \int_{-\infty}^{+\infty} z\rho_m(z)dz = -4\pi\sigma Z_m$$

At the uncharged metal/vacuum interface where the charge distribution is $\rho_m^0(z)$, $\chi^m(\sigma)$ corresponds to the dipole barrier χ^m at the metal surface, referred to as the "surface potential of the metal" or "electron overlap potential" in the electrochemical literature. Note that, given the sign. convention adopted, χ^m equals the electrostatic potential inside the metal minus the potential outside. At the metal/solution interface, $\rho_m(z)$ differs from $\rho_m^0(z)$ as a consequence of the metal–solution coupling. Thus, $\rho_m(z)$ reflects the perturbation due to the presence of the solution and the fact that it is possibly charged. The two effects may be formally considered separately by defining first the charge density $\Delta\rho_m(z) = \rho_m(z, \rho) - \rho_m(z, \sigma, = 0)$. This quantity accounts for the change in the local density of charges in the metal in the presence of the solution when the interface is charged. Second, one may consider the total variation $\delta\rho_m(z)$ of $\rho_m(z, \rho)$ with respect to the uncharged metal in vacuum: $\delta\rho_m(z) \equiv \rho_m(z, s) - \rho_m^0(z)$. Then we may write

$$\chi^m(\sigma) = \chi^m + \delta\chi^m(\sigma) \tag{7a}$$

$$\delta\chi^m(\sigma) = -4\pi \int_{-\infty}^{+\infty} dz\, z\delta\rho_m(z) \tag{7b}$$

or, equivalently,

$$\chi^m(\sigma) = \chi^m(\sigma = 0) + \Delta\chi^m(\sigma) \tag{8a}$$

$$\Delta\chi^m(\sigma) = -4\pi \int_{-\infty}^{+\infty} dz\, z\Delta\rho_m(z) \tag{8b}$$

where, by definition, $\Delta\chi^m(0) = 0$, in contrast to $\delta\chi^m(0)$. From the definitions in Eqs. (7) and (8), we see that the "surface potential" $\chi^m(\sigma=0)$ for the uncharged metal in the presence of the solution differs from the "surface potential" χ^m for the uncharged metal in vacuum. Accordingly, we may write

$$\Delta\chi^m(\sigma) = -4\pi\sigma Z_0(\sigma) \tag{9}$$

where $Z_0(\sigma)$ has the dimensions of length and is possibly non-vanishing at the pzc. By using Eqs. (9) and (8b), $Z_0(\sigma)$ can be written as

$$Z_0(\sigma) = (1/\sigma) \int dz\, z\Delta\rho_m(z) \tag{10a}$$

From the definition of $\Delta\rho_m(z)$, $Z_0(\sigma)$ given in Eq. (10a) is the equivalent at the metal/solution interface of the "center of mass" or "centroid" of the charge that an external electrical held induces on the metal surface for the case considered in the literature of the metal/vacuum interface. In that case, assuming that the distribution of the metal ions in unaffected by the field, $\Delta\rho_m(z)$ in Eq. (10a) must be replaced by $\delta n(z) = n_{\sigma=0}(z) - n_s(z)$, where, following the conventions used, ri,(z) is the number density of electrons at a position z when the metal surface carries a charge o–per unit area and when atomic units ($e = 1$) are used for the electronic charge. It will also be convenient to define the differential quantity $X_0(\sigma)$:

$$X_0(\sigma) = \partial[\sigma Z_0(\sigma)]/\partial\sigma \tag{10b}$$

which can also be written as

$$X_0(\sigma) = -(4\pi)^{-1} \partial X^m(\sigma)/\partial\sigma \tag{10c}$$

Equation (10c) is obtained from Eq. (10b) by using Eq. (9) for $\Delta X^m(\sigma)$ and Eq. (8a) for $X^m(\sigma)$. In Eq. (10c), $X_0(\sigma)$, which has the dimensions of length, measures the rate of change with the surface charge of the "surface potential" of the metal. The physical interpretation of the length $X_0(\sigma)$ can be better seen from Eq. (10b), if the derivative is replaced by the ratio of infinitesimal variations and Eq. (10a) for $Z_0(\sigma)$ is used:

$$X_0(\sigma) = \int dz\, z[\rho_m(z, \sigma + \delta\sigma) - \rho_m(z, \sigma)]/\delta\sigma \tag{10d}$$

In Eq. (10d) the infinitesimal charge $\delta\sigma = \int dz\, [\rho_m(z, \sigma + \delta\sigma) - \rho_m(z, \sigma)]$ is related to a redistribution of the metal electrons across the interfacial region. Thus, $X_0(\sigma)$ has the meaning of the "center of mass" of the spatial distribution of the local excess or deficit of electrons corresponding, for example, to an infinitesimal variation of the applied potential. We may now define a similar quantity $X_s(\sigma)$ for the solution, with ρ_m being replaced by ρ_s in Eqs. (6)–(10). Then the capacitance may be alternatively written as

$$1/C = 4\pi[X_s(\sigma) - X_0(\sigma)] - 4\pi \, \partial / \partial\sigma \int_{-\infty}^{+\infty} P(z) \, dz \qquad (11)$$

It is recalled here that $P(z)$ is the molecular polarization. This important relation will be the starting point of our theoretical analysis of the capacitance. It is important to note that, as a consequence of the coupling between the two sides of the interface, the different terms in this expression cannot be calculated from the properties of each side taken separately. The approximations used for these terms precisely define different models for the capacitance.

From Eqs. (4) and (9), another equivalent expression for $\Delta\phi_s^m$ can be obtained. By using Eq. (8a) for $X^m(\sigma)$, $\Delta\phi_s^m$ can be written as

$$\Delta\phi_s^m = X^m(\sigma = 0) + 4\pi\sigma \, (Z_s - Z_0) - 4, \int_{-\infty}^{+\infty} P(z) \, dz \qquad (12)$$

In Eq. (12) the contribution to $\Delta\phi_s^m$ of the surface potential $X^m(\sigma = 0)$ of the uncharged metal in the presence of the solution has been isolated.

Analysis of Classical Models

In order to understand the structure of classical models in the light of the previous analysis, we compare, Eq. (12) to the classical expression for the potential drop. Following the classical splitting, $\Delta\phi_s^m$ is written as

$$\Delta\phi_s^m = g_{dip}^m + g_s^m \text{(ion)} - g_{dip}^s \qquad (13)$$

In this expression, g_{dip}^m, g_s^m(ion), and g_{dip}^s are the potential differences due to the surface dipole of the metal, the free charges at constant dipole orientation, and the orientation of the permanent dipoles of the solvent molecules at the interface, respectively. This equation represents a particular separation of contributions to the potential drop $\Delta\phi_s^m$ and is based on the separation of the molecular polarization into a contribution related to the orientation of the permanent dipoles, p^{or}, and a contribution related to their polarizability, p^{pol}. The link between Eqs. (13) and (12) has been discussed in Ref. 5. Here, we proceed a little further. By definition, g_s^m(ion) vanishes at the pzc. Then, if we assume such a separation of $P(z)$ in Eq. (12), we may identify $\left[g_{dip}^m\right]_{\sigma=0}$ with $X^m(\sigma=0) = X^m + \delta X^m(0)$ and g_{dip}^s with $4\pi \int_{-\infty}^{+\infty} P^{or}(z)\,dz$. At a charged interface the comparison of the various expressions of $\Delta\phi_s^m$ with the particular separation in Eq. (13) is less straightforward because all the terms depend on charge. There is no problem with g_{dip}^s and $4\pi \int_{-\infty}^{+\infty} P^{or}(z)\,dz$ since both arise from the permanent dipoles. The status of $g_{dip}^m + g_s^m$ (ion) in Eq. (13) at the charged interface depends in the traditional approach on the models of the inner layer. It is generally considered 311 that these terms give rise to a linear potential drop (σ/K_{ion} + constant). In the simplest models, the inner layer in the absence of specific ionic adsorption consists of a monolayer of water located between the electrode plane and the outer Helm–holtz plane. Then K_{ion} is the capacitance of a capacitor consisting of a dielectric medium of thickness e and dielectric constant E between two fixed charged planes: $1/K_{ion} = 4\pi e/\varepsilon$. The dielectric constant ε is related to the distortional polarization of the solvent, assumed to be a linear function of the interfacial field. It, as usual, the "thickness" e of the capacitor is taken as equal to one molecular diameter, then from the value of the inner–layer capacitance at sufficiently negative charge, ε is approximately equal to 6. Note that in the model of Bockris and co-workers which was intended for the study of specific ionic adsorption, two water layers with different dielectric constants are considered as forming the inner layer. The important point here is that following the classical analysis the contribution of g_{dip}^{m} and g_{s}^{m} (ion) to the capacitance is a constant.

The formally exact expression for the potential drop given by Eq. (12) clarifies the approximations involved in the classical models. From Eq. (12) we have the following identification:

$$g_{dip}^{m} + g_{s}^{m} \text{ (ion)} = X^{m}(\sigma = 0) + 4\pi\sigma(Z_s - Z_0) - 4\pi \int_{-\infty}^{+\infty} P^{pol}(z)\, dz$$

In these models Z_0 is the position of the "electrode plane," considered as fixed, that is, the geometrical boundary of the metal. The charge σ is assumed to be localized on the plane $z = Z_0$; that is, $\delta\rho_m(z)$ has the form $\delta\rho_m(z) = \sigma\delta(z - Z_0)$, where δ is the Dirac delta function. In actual models, Z_0 is taken as the origin of the coordinates ($Z_0 = 0$). Similarly, at high enough electrolyte concentrations and in the absence of specific ionic adsorption, the excess charge $\delta\rho(z)$ in the solution is assumed to be distributed on the outer Helmholtz plane located at Z_{oHp}: $\delta\rho_s(z) = -\sigma\delta(z - Z_{oHp})$. Then $(Z_s - Z_0)$ in Eq. (12) reduces to the geometrical distance $(Z_{oHp} - Z_0) = e$, considered as the "thickness" of the inner layer. Apart from the dielectric continuum in the diffuse layer, the molecular polarization $P(z)$ in these models is localized in the monolayer adjacent to the electrode. When the linear distortional polarization in this monolayer is taken into account, the quantity $(Z_{oHp} - Z_0)$ is divided by ε.

The subsequent linear potentia drop gives the classical expression of the capacitance K_{ion}. Obviously, in this case the "thickness" of the inner layer is charge independent. However, in their well-known paper, MacDonald and Barlow considered the possibility of a variation of the thickness of the inner layer with charge. Note that this aspect has a central importance even in the most recent versions of the traditional approach as it appears in the work of Guidelli and Aloisi.

If we go back to Fq. (12), the results of' the calculations show that in contrast to the assumption made in the classical models, $4\pi\sigma(Z_s - Z_0)$ is not a linear function of charge since both Z_s and Z_0 depend on σ. Although the potential drop $4\pi\sigma(Z_s - Z_0)$ is formally the same as that for two oppositely charged layers located at Z_s and Z_0, the distance $(Z_s - Z_0)$, that is, the "thickness" of the capacitor, is not constant with the charge. We stress that in Eq. (12), Z_s and Z_0 are not physical positions, but correspond to the centroids of charges distributed across the interface, which eventually change with the interfacial field. Physically, Z_0 being very small and independent of the interfacial field means that the induced charge on the metal surface has no spatial extension. This also means that an external electrical field does not penetrate the metal, being completely screened beyond Z_0. Although field penetration was conceptually accepted, this effect was, with very few exceptions (see below), considered as negligible in the calculation of the capacitance. In its turn, Z_s being constant means that the ions in the outer Helmholtz plane approach the metal surface always at the same distance, irrespective of the strength of the interfacial field. This means that the geometrical plane simulating "the surface" acts as a sharp boundary for the solvated cations.

The main task in the classical approach is then to determine the orientational contribution P^{or} to the polarization $P(z)$ in the layer adjacent to the electrode from a molecular model. It is there that the nature of the electrode enters the classical models. The role of the metal is then essentially indirect and appears mostly through a chemical interaction giving rise to some preferred orientations of the molecules. This aspect of the models has been extensively discussed in the literature. The reader is referred to Refs. 1, 2, and 6, for example. This chemical interaction may explain the shape of the curves and their variation with the nature of the metal (compare, for example, the work on the Hg/NaF interface by Parsons and that of Valette on silver, both of whom used the same model). Recall also the discussion of Trasatti and of Parsons on the relation between the strength of the metal–solvent interaction and the magnitude of the capacitance.

The interaction of the metal surface with the solution has also been discussed in terms of physical interactions. Electrostatic interactions including image forces on the ions or dipoles together with dispersion forces (in their classical formulation) were introduced in models of various sophistication, such as in the work of Watts–Tobin Bockris and co-workers Barlow and MacDonald, Levine et al., Fawcett et al., Damaskin and Frumkin, Parsons, and Guidelli and Alosis. A vast amount of work has been done by these authors and many others in order to explain various experimental aspects, generally with appreciable success' However, we will now point out some aspects where this description of the role of the metal may not be sufficient.

Why a New Description of the Role of the Metal?

Despite the apparent success mentioned above, the classical models may be criticized on two different levels. First, these models use parameters which either are adjustable or sometimes correspond to quantities which are ill defined on a microscopic level. Just as an example of this last aspect, we mention the arbitrary values of the "dielectric constant" of the two water layers in the model of Bockris and co-workers or' the value of the ratio of the dipole moment of the clusters to that of the monomer in the work of' Parsons. The observed sensitivity of the results to the precise value of some of these parameters is a real drawback of these models. Second, except for very general considerations regarding the contribution of the electron overlap, with no treatment of it on a microscopic level, all these models do not take into account the new description of the electrode surface and its consequences for the analysis of' properties as important as the capacitance.

Progress in the understanding of' the electronic properties of metal surfaces which followed to a large extent from the work of Lang and Kohn indeed provides answers to some of the questions raised by the use of the (over)simplified description of the electrode surface in the traditional approach. Just above, three assumptions which are implicit in the classical description of the electrode were mentioned. Among the problems encountered in the classical analysis of the interface is the fact that $X^m(\sigma)$ is considered differently depending on the state of charge of the interface. At the pzc, it is widely accepted that the presence of the solution may induce a rearrangement in the electronic distribution giving rise to $\delta X^m(0)$ in Eq. (7a). However, at a charged interface, the excess charge $\Delta \rho_m(z)$ in Eq. (8b) induced by an external field is assumed to be distributed on a geometrical plane whose position Z_0 is charge-independent. In other words, in the classical models, the electronic structure of the metal is sensitive to the presence of molecules but is practically not affected by an external field. Although from obvious physical considerations a possible effect of the applied field on the response of the electrons in the metal was accepted, this effect was considered as negligible or was not taken into account explicitly. In fact, the assumption that the induced charge $\delta\rho_m(z)$ is sharply peaked at the geometrical surface is not correct: it is now recognized that $\delta\rho_m(z)$ has a more diffuse character. This means, for example, that there is an appreciable penetration of the electric field into the metal. As recalled above, this aspect of metal "nonideality" is in fact not a new concept. Indeed, it has been recognized since 1928 that the effect of field penetration at a metal surface may play a major role in the capacitance, eventually leading to what is sometimes referred to as the "Rice paradox" (see, for example, Refs. 154 and 170). Indeed, it follows from the estimation of the effect of field penetration made by. Rice that the metal contribution gives the entire value of the capacitance, with little place for a contribution from the solution. The difficulties arise largely from the fact that in Rice's model the metal electrons were subject to an infinite potential barrier near the surface. It was only in the beginning of the seventies that a self-consistent treatment of the electron gas subject to a realistic potential barrier at a metal surface became successful. This point has been taken into account in the models proposed in the early eighties for the metal/electrolyte interface, which removed the difficulties of Rice's model. The relevant work will be discussed in the last two sections of this chapter. See also the chapter by Goodisman in this series.

The modern description of the structure of the metal surface definitely clarifies the notion of "electrode plane." As shown in 1973 by Lang and Kohn, in the case of the response of the metal to a weak field, the electrode plane is to be identified with the image plane, whose position X_0 is given by the centroid of the charge induced by a weak uniform field [see Eq. (10d) and Fig. 10]. This position X_0 of the "electrode plane" is approximately fixed only in the linear regime close to the pzc. Away from the pzc, the shape of $\Delta\rho_m(z)$ changes noticeably with the value of the net (and non-infinitesimal) charge σ. Its center of mass Z_0 [the integral of $\Delta\rho_m(z)$ is σ] is also charge-dependent. In the limit of weak surface charges, X_0 and Z_0 are of course identical. We shall see in the next section that X_0 (or Z_0) can be determined unambiguously with respect to the last ionic plane of the solid (and can also be determined from theory in the case of a liquid, at least in principle). For all electrical properties, this position X_0 should be considered as the *effective location* of the metal surface. This point is relevant to electrochemistry in many respects, and part of this chapter will be concerned with this point and related aspects. Let us already mention some important points. Its relevance in the determination of the capacitance can be visualized if one considers the metal in the presence of an assumed ideal charged plane located at a distance D far from the geometrical surface of the metal. The whole system is equivalent to a capacitor whose capacitance is given by $C^{-1} = 4\pi(D - X_0)$ rather than by the classical expression $C^{-1} = 4\pi D$. This aspect will be discussed in detail in Section III.4 (see Fig. 8).

More generally, a description of the metal surface which goes beyond its representation as a geometrical plane may be relevant in the estimation of interaction energies. For example, image forces in their classical formulation are common ingredients in many classical models. It is then important to recall that the classical formulation of image forces is incorrect near the surface. In classical electrostatics, the image force on a point charge at a distance Z from an ideal conductor ($Z = 0$ is the geometrical boundary of the metal) varies as $1/Z$. A microscopic description of the metal leads actually to a slower variation at small values of Z. A similar situation occurs for dispersion forces (see Section III). Since typical Values of Z for ions or dipoles in contact with the metal are of the order of a few angstroms, the correction to the classical expression is indeed substantial. This may have serious consequences for the predictions of models where electrostatic forces (or dispersion forces) play an important role in the structuring of the particles in contact with the electrode. Then the consideration of the real structure of the electrode may be important for an accurate description of the interaction of adsorbed particles with the surface. For example, it is implicit in the classical view that the electrode acts as a sharp boundary representing the short-range surface–adsorbate repulsion. The actual metal–adsorbate interaction potential exhibits in fact a smoother variation with distance. This would mean that there is no reason to locate the position of the center of the water layer adjacent to the electrode a priori at one molecular radius from the "metal surface." A similar remark holds also for the position of the outer Helmholtz plane.

To sum up, a characteristic distance of the interface–the effective position of the electrode surface–can be determined now from the physical properties of the metal. Other important distances such as the distance of closest approach of the solvent molecules or the ions to the electrode should also be determined from the knowledge of their actual interaction with the metal surface by using some appropriate criteria (this aspect is discussed in Section IV.2). We thus expect a detailed description of the metal surface together with its interaction with adsorbed species to be a necessary step for a satisfactory description of the role of the metal in interfacial properties.

Before going further in this direction, a few words should be said about the description of the solution side of the interface. This field of investigation, which has been renewed in the last decade, will not be treated in this chapter (the interested reader is referred to the reviews by Carnie and Torrie and by Henderson). The major task is to calculate $P(z)$ and $\rho_s(z)$ in Eq. (2) for a given model of the solution, by using the tools of the modern statistical mechanical theory of inhomogeneous liquids (integral equations, for example). Some interesting results have already been obtained for model systems. However, they hove not yet been extended to real situations. An important aspect is that in most of these works the metal is replaced by a charged wall, which means $Z_0 = 0$. This group of methods includes computer simulations. We note here the interesting results obtained recently by Heinzinger and Spohr. See also references in their review paper. The development of such techniques will certainly bring valuable information on the properties of the solution in the near future.

An Alternative to a Fully Microscopic Approach

We may now summarize the situation: the traditional approach has brought a large wealth of information, but also has its limitations and has left many questions open. The modern statistical mechanical description of the solution based on models which go beyond the "primitive" model of point ions in a dielectric continuum is still in a preliminary stage of development. The work of the last ten years based on a microscopic description of the metal seems more directly related to experimental aspects. However, all recent attempts toward an *ab initio* calculation of the capacitance which couple the recent view of the metal to models of the solution beyond the "primitive" model have failed to reproduce the experimental data, at least insofar as the charge–capacitance curves and their variation with the nature of the metal are concerned. In our opinion, this situation is understandable: besides the well-known limitations of the statistical mechanical methods presently available, such as the mean spherical approximation (MSA) or its variants, the direct calculation of the capacitance requires that each term in Eq. (11) should be calculated with high accuracy since the inverse capacitance results from cancellation between terms of comparable magnitude. We hope that Section VII, devoted to our own work, will convince the reader on this point.

An alternative to the direct calculation of the capacitance by the evaluation of all terms in Eq. (11) may consist in calculating some of them as accurately as possible. Then if the calculated terms are separated from the experimental value of the (inverse) capacitance, a semiempirical estimate of the *remaining contributions* is deduced. Then one may have some hope that a simple interpretation of these terms which contain more limited information can be found *a posteriori*. In our recent work we found that this semiempirical approach proved to be rather fruitful. Practically, we define from Eq. (11) a contribution C_m. By using the experimental C_i, we isolate a contribution C_s by using the relation $1/C_s = 1/C_i - 1/C_m$. The precise meaning of C_m and C_s will be discussed in detail in Section VII.2. We just indicate here that the calculation of C_m requires a model for the metal and for the metal–solution coupling. A review of the theory of metal surfaces and metal–adsorbate interactions will be the subject of the next section.

2

Electronic Properties of the Metal Surface

The preceding sections have shown that some properties of the metal/solution interface are determined to some extent by physical properties of the metal side. For example, we have seen that for mercury or gallium electrodes, the interfacial tension is determined largely by the surface tension of the metal/vacuum interface (Section II.1). Also, the potential of zero charge has been known for a very long time 61 to be related to the value of the electronic work function Φ. At a charged interface, a question of fundamental interest for the theory is to determine bow these metal properties are affected by the interfacial field. A natural requirement that a model of the metal at the interface should fulfill is that it must first be satisfactory for typical properties of the bare surface such as the surface tension γ or the work function Φ. In the case of weak interfacial coupling, the models fulfilling this criterion may be used as a good starting point in the modeling of the interface.

From the experimental point of view, the surface tension is a more characteristic property of the metal than the work function. Indeed, while q) varies roughly by a factor of two in the periodic table, the surface tension of a light alkali such as Cs (40 dyn/cm) is an order of magnitude smaller than that of Hg (400 dyn/cm) or Al (1000 dyn/cm) and 50 times less than that of Fe (2000 dyn/cm). The surface tension should then provide a much more stringent test of theory than the work function.

Unfortunately, except in the case of the alkalies and a few polyvalent metals, all models proposed so far (apart from semi–phenomenological formulas) for a microscopic derivation of the surface tension for sp metals have failed to reproduce the experimental values of γ, especially for liquid Hg (see, for example, the recent work of Hasegawa 63 and references therein). For these metals, the difficulties originate to a large extent from the fact that an important contribution due to conduction electrons competes equally with that related to the ions. This cancellation makes the theoretical results very inaccurate. This in contrast with the case of the transition metals, which will not be discussed here; for these metals, only one phenomenon dominates, namely, the filling of the d band. However, if only the variation of γ with the potential, which ultimately determines the capacitance through the Lippmann equation, is required, the actual magnitude of the surface tension might not be relevant. For this purpose, what seems important is to have a good description of electrostatic contributions to γ. The electrostatic energy depends on the average distribution of charges at the surface. In addition, a correct description of the electrostatic potential at the surface is also required for a satisfactory description of the work function. Thus, while a good account of electrostatics seems necessary for both γ and Φ, an accurate calculation of the magnitude of γ is not strictly required. This point is further illustrated by the following example.

For the uncharged surface, it is well known from the theoretical work of Lang and Kohn that, except in the case of the alkalies, the jellium model, in which the discrete nature of the ions is ignored, predicts unreasonable negative values of γ (more precisely the surface energy U_s) while the values of Φ given by the same model are not in significant error when compared to experiment. The introduction of the discrete nature of the ions as a perturbation changes dramatically the values of U_s, which become much closer to the experimental values, while the values of Φ remain roughly in the same range. Much of the improvement is due to the appearance of additional terms in the expression of the surface energy. Among these new terms, the cleavage energy of the solid makes a very large contribution to U_s. However, this term is only related to the properties of the ionic lattice. Another effect comes from the change in energy related to the noncoulombic nature of the short-range interaction between the electrons and the ions; this term is calculated with the electronic profile corresponding to the Jellium model in Ref. 47. These terms are important for determining the magnitude of U_s, but they do not dramatically change the values of Φ, that is, the potential drop at the interface. This ultimately means that a model which does not predict accurate values for the energy may still give reasonable values of Φ.

One may ask whether similar behavior occurs also in the case of a charged surface. For solids, the ions will contribute to the energy through terms not affected by an external electric field, since the latter is almost completely screened by the conduction electrons at the position of the ions. The situation is then similar to that existing at the pzc. Then it will be sufficient to discuss a possible effect of the field related only to nonzero ionic size (in the electron–ion interaction energy) on the work function.

Due to the diffuse nature of the ionic profile in the case of liquid metals, screening of external field,, by the free electrons might be less efficient than for solids. The ionic profile may then vary with charge, with a possible effect on both γ and X^m. In this case, no definite conclusion can be reached prior to calculations.

To summarize, at the pzc and for solids it seems that reasonable values of the work function may coexist within the same model with somewhat less satisfactory values of the surface energy. At a charged surface, only the correct behavior of the electrostatic potential is needed. For all these reasons, in the following we will mainly present values of the work function in the case of solids. The case of liquids will be discussed only on a qualitative level.

Density Functional Theory

The density functional theory and its applications to the description of bulk and surface properties of metals have been the subject of numerous and extensive reviews. The interested reader is referred to Refs. 62. 65, and 66 for a detailed presentation of the theory. For a more specific application to electrochemistry, see, for example, Refs. 67 and 68 and the chapter by Goodisman in this series. Here, only the aspects of the theory which will be useful for the discussion of recent models of the metal/electrolyte solution interface will be considered.

(i) Ground State Energy of the Electron Gas

The theoretical description of the electronic properties of metals requires the solution of a complicated quantum–mechanical many–body problem. Forsp metals, including the alkalies and some polyvalent metals (e.g., Al, Mg), the electronic states in the metal can be separated into core states and conduction band states. For most properties of interest, one has to consider explicitly only states in the conduction band. Then, one is led to calculate the properties of a gas of conduction electrons immersed in a lattice of ions (the core electrons and the nucleus). It can be shown that the energy of these metals in the bulk phase can be calculated by considering the conduction electrons as nearly free electrons, weakly perturbed by the presence of' a lattice of ions. This description is based on the so-called pseudo-potential theory. In this approach, the electronic wave functions behave as plane waves outside the ionic core, while they are expanded in terms of core state functions near the nucleus. This leads to the replacement of the strong lattice potential by a weak pseudo-potential operator, which allows the use of perturbation theory. The description of bulk properties takes then great advantage of translational invariance and of the possibility to using, in general, very simple models for the pseudo-potentials.

In the case of surface properties, the problem is further complicated by the loss of translational invariance introduced by the presence of a surface, in the vicinity of which the gas of conduction electrons is strongly inhomogeneous. This problem has been made tractable by Hohenberg and Kohn and Kohn and Sham, who introduced the density functional formalism (DFF). This formalism is specially devised for the study of strongly inhomogeneous systems like the electron cloud of atoms or molecules or the inhomogeneous electron gas appearing when the surface of a solid is created. The main idea involved in the density functional formalism is that the problem should be greatly simplified if, instead of trying to determine the complex many-body wave function of the system, one focuses on the electronic density $n(\mathbf{r})$ at a given point $\mathbf{r}$ of the interface. Using the density as the basic variable, one constructs an appropriate functional $E[n(\mathbf{r})]$ which, when evaluated with the exact density, will correspond to the ground state energy. A fundamental property of this formalism is that the ground state energy of the electron gas in the presence of an external potential $V(\mathbf{r})$ is the minimum value, with respect to variations of $n(\mathbf{r})$ at a constant number of electrons, of $E[n(\mathbf{r})]$. The functional $E[n(\mathbf{r})]$ can be written as:

$$E[n(\mathbf{r})] = G[n(\mathbf{r})] + \tfrac{1}{2} \int d\mathbf{r}\, d\mathbf{r}'\, n(\mathbf{r})n(\mathbf{r}')/|\mathbf{r}-\mathbf{r}'|$$
$$+ \int d\mathbf{r}\, n(\mathbf{r})V(\mathbf{r}) \tag{14}$$

where $G[n(\mathbf{r})]$ contains the kinetic energy of the electrons, the exchange energy related to the Pauli principle, and that arising from the correlated motion of the electrons. The second term is the electrostatic self–energy of the electron gas, and the last term is its interaction energy with the external potential $V(\mathbf{r})$, which may correspond, for example, to the potential of the semi–infinite lattice of ions. Note that one should add to Eq. (14) some constant terms independent of $n(\mathbf{r})$ in order to deal with finite quantities.

For a particular application, the main task is to devise some approximation for the unknown functional $G[n(\mathbf{r})]$, which contains all the complexity of the many–body problem. One major ingredient in the theory is the so-called "local density approximation" (LDA) for the exchange-correlation part, E_{xc}, of $G[n(\mathbf{r})]$:

$$E_{xc}[n(\mathbf{r})] = \int d\mathbf{r}\, n(\mathbf{r})\{e_x[n(\mathbf{r})] + e_c[n(\mathbf{r})]\} \tag{15}$$

where e_x and e_c are, respectively, the exchange and correlation energies per particle of a *uniform* electron gas of density $n(\mathbf{r})$. The idea involved in the LDA is to approximate the local properties of the electron gas by those of a gas of uniform density equal to the local density $n(\mathbf{r})$. While nonlocal treatments of E_{xc} have been proposed (see, for example, references in Ref. 66), most applications of the formalism are within the LDA. Frequently used expressions for exchange–correlation energies are the Hartree approximation for e_x: $e_x(n) = -3/4(3n/\pi)^{1/3} = -0.458/r_s$ for exchange; the Wigner interpolation formula for e_c: $e_c(n) = -0.44/(r_s + 7.8)$; or the analytically convenient form due to Pines and Nozieres: $e_c(n) = \tfrac{1}{2}[-0.0115 + \ln(r_s)]$. The parameter rs, defined by $r_s = (3/4\pi n)^{1/3}$ where n is the bulk electron density, is a convenient way of characterizing the average free electron density of the homogeneous metal: r_s is the radius of the sphere containing on average, one electron.

In addition to the necessary approximations for $G[n(r)]$, the effect of the semi–infinite lattice of ions cannot be treated exactly. One is then forced to construct some model of the surface in order to make the problem tractable. Some methods proposed for treating the functional $G[n(\mathbf{r})]$ and the related models of the surface are briefly presented below.

The Jellium Model

The simplest possible representation of the surface is found in the so–called "jelliurn" model. In this model, the discrete nature of the semi–infinite lattice of ions is ignored. Rather, it is replaced by a structureless uniform background of positive charge, terminating abruptly at the "surface" (Fig. 3) and which neutralizes the total negative charge of the electron gas. In this picture, the problem is reduced to that of an interacting electron gas subject to the coulombic potential $V(z)$ due to the semi–infinite jellium. Due to the structureless nature of the latter, the properties of the electron–jellium system are invariant in the direction parallel to the surface. This introduces a major simplification in that the problem is now reduced to a one–dimensional one. This feature is the reason for the wide popularity of the jellium model.

An approximate treatment of this problem Was proposed by Smith in 1969. A more exact treatment was published by Lang and Kohn in 1970.Most of the subsequent. developments in the theory of metal surfaces followed from the work of these authors. Since both the method proposed by Smith and the self–consistent calculations of Lang and Kohn (LK) have been used at the metal/solution interface, we briefly describe the two methods, Details can be found in Ref. 5.

he LK approach is based on the fact that the minimization of $E[n(\mathbf{r})]$ given in Eq. (14) is equivalent to finding the solution of a Schrodinger–like equation for electrons in an effective potential V_{eff}:

$$\{(-\hbar^2/2m)\Delta + V_{eff}[(\mathbf{r}), \mathbf{r}]\} \psi_1 = \varepsilon_1 \psi_1 \quad (16)$$

where the first term on the left–hand side is the usual kinetic energy operator. In this equation, V_{eff} is the sum of $V(\mathbf{r})$, the coulomb potential of the electrons, and the exchange–correlation potential $\delta E_{xc}[n(\mathbf{r})]/\delta n(\mathbf{r})$, which in the LDA is equal to $d[n\varepsilon_{xc}(n)]/dn$ where $\varepsilon_{xc}(n)$ is simply $e_x[n(\mathbf{r})] + e_c[n(\mathbf{r})]$ as discussed above in Section II1.2(i). The effective potential depends on the electron density $n(\mathbf{r}) = \Sigma_l \psi_l^*(\mathbf{r})\psi_l(\mathbf{r})$, so that the equations must be solved self–consistently. This procedure is similar to that followed in a Hartree–Fock calculation, with which the reader may be more familiar. The difference is that while Hartree–Fock treats exchange exactly but ignores correlations, DFF + LDA considers both in an approximate way.

The approach of Smith is much simpler than that of LK in that it avoids having to find the complicated solution of Eq. (16) and the self–consistent calculation of V_{eff}. Instead, the functional $E[n(z)]$ is directly minimized in a given class of trial functions for the density profile $n(z)$. These functions are chosen in such a way as to reproduce the gross features of the exact profiles as obtained from an LK calculation. They depend on a set of variational parameters $\{\alpha_i\}$. Some of these are determined by general conditions of continuity of $n(z)$ and its derivative and by the requirement of charge conservation. The remaining free parameters are determined by the minimization of $E[n(z)]$. The practical

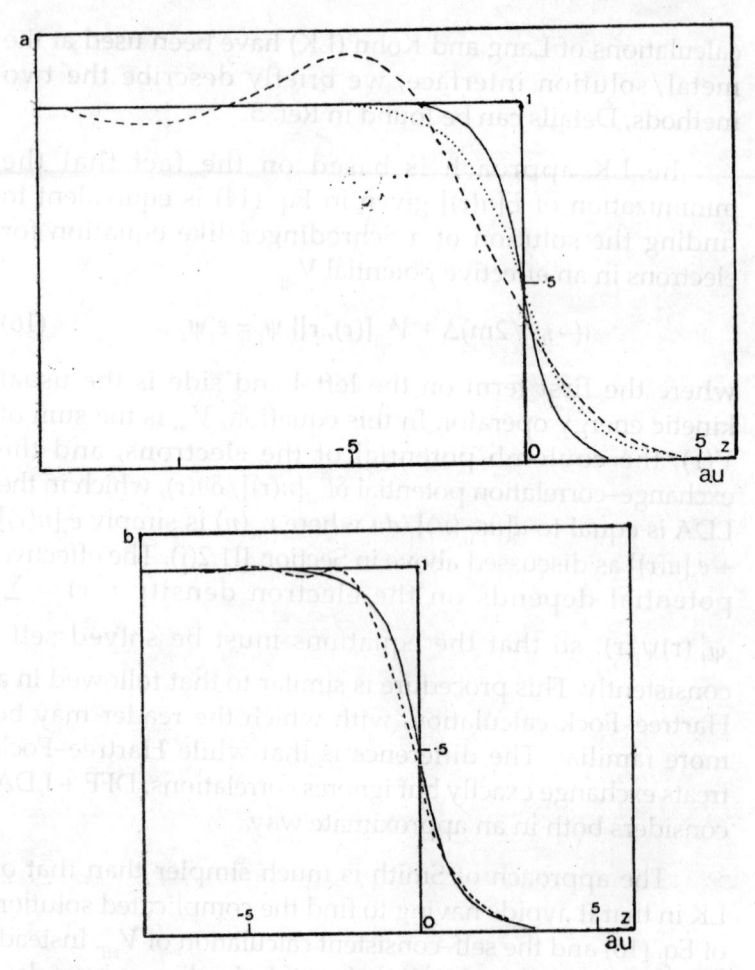

Figure 4. Calculated density profiles $n(z)/n$ for $r_s = 6$ (a) and $r_s = 2$ (b) in the jellium model. (–) from Smith; (- - -) from Lang and Kohn. In (a), (. . .) is from the work of Ma and Sahni and shows the effect of second gradient terms.

procedure consists in minimizing the surface energy U_s per unit area A, obtained by subtracting from $E[n(\mathbf{r})]$ the contribution of an electron–jellium system whose electron density is uniform up to the jellium edge:

$$U_s = \{E[n(z)] - E(n)\}/A, \qquad \partial U_s/\partial \alpha_i = 0 \qquad (17)$$

Note that the approach of Smith also uses the LDA for the exchange–correlation energy. An expansion of the kinetic energy $T[n(z)]$ in terms of density gradients is further made, in contrast to the approach of LK, which treats this quantity exactly. Then, $T[n(z)]$ is written as

$$T[n(z)] \approx \int dz\, n(z)\{t_0[n(z)] + t_1[n(z)] + t_2[n(z)] + \ldots\} \qquad (18)$$

The first term in the integrand corresponds to a simple Thomas–Fermi approximation. Developments up to second gradient terms are now commonly used, following the work of Ma and Sahni, who pointed out their importance.

As an illustration, we have plotted in Fig. 4 the density profiles obtained by the two methods for $r_s = 6$ and $r_s = 2$, corresponding roughly to Cs and Al. The Smith profiles, obtained with an exponential type of trial function, are monotonic while in the exact profiles the Friedel oscillations are clearly visible in the case of Cs. Note that these oscillations are less marked when the bulk electron density n increases or equivalently when the value of rs decreases. In general, the method of Smith is less accurate than that of LK, and the profiles are only approximations of the exact ones. This may have an important consequence in the calculation of the electrostatic potential. The latter can be in significant error when calculated directly by using in Eq. (6) the profiles obtained by the minimization of U_s. However, this feature can be circumvented by using an alternative determination of the potential, as detailed below. If such a procedure is followed, one can really take advantage of the simplicity of the method of Smith, compared to that of LK.

Calculation of the Work Function. Position of the Image Plane

By definition, the work function Φ is the minimum

work required to extract from the metal one electron at the Fermi level and transfer it to the vacuum. Then it follows from the general formalism that Φ is given by

$$\Phi = X^m - \bar{\mu}_e \qquad (19)$$

where X^m is the electrostatic energy barrier at the metal surface, and $\bar{\mu}_e$ is the bulk chemical potential of the electrons relative to the mean electrostatic potential in the metal interior and is calculated from the theory. Note that since atomic units for the electronic charge are used in Eq. (19), X^m is also the surface dipole barrier as defined in Eq. (6). In the jellium model, the charge density is $\rho_m(z) = n\theta(-z) - n(z)$, where $n\theta(-z)$ is the background density, $\theta(-z)$ being the Heaviside step function (the metal occupies the $z \leq 0$ half–space). Then from Eq. (6) we have:

$$X^m = 4\pi \int dz\, z[n(z) - n\theta(-z)] \qquad (20)$$

Equation (19) is sometimes referred as the "Koopman's theorem" expression of the work function. Note that Eq. (19) is the same as the expression used in the electrochemical literature. Equation (19) shows that Φ results from a cancellation between the surface term X^m and the bulk contribution $\bar{\mu}_e$. Especially in the case of low r_s, both terms of Eq. (19) must be calculated very accurately if this Koopman's theorem expression of Φ is used. As an extreme example, the recent work of Perdew and Wang has shown that for very low r_s, values of Φ of a few electron volts are obtained from values of X^m and $\bar{\mu}_e$ of more than 150 eV. In the case of a jellium density corresponding to Al, $\Phi = 3.78$ eV while $X^m = 6.59$ eV. If the desired accuracy is a few tenths of an electron volt, then the value of Φ in Eq. (19) obtained with X^m calculated in Eq. (20) with the approximated profile $n(z)$ may not be accurate enough. If one does not use the LK scheme, this problem can still be avoided by using the alternative expression of the work function called the "change in self-consistent field" expression of Φ, denoted $\Phi_{\Delta SCF}$:

$$\Phi_{\Delta SCF} = dU_s/d\Sigma \,|_{\Sigma=0} \qquad (21)$$

where Σ is the number of electrons per unit area on the surface. This less familiar expression of Φ is a consequence of the general definition of the work function. It may be understood as the change in energy (per unit area) of the metal when an infinitesimal charge is induced on its surface, due to the transfer of a few electrons from the metal to the vacuum. Since the induced charge is localized on the surface, this change in energy results from a change in surface energy. Because of the stationary property of $E[n(z)]$ (viz. U_s) with respect to a variation of the density profile, the ΔSCF expression is much less sensitive to errors in the profile than the direct Koopman's theorem expression. Note that, in practice, one may obtain $\Phi_{\Delta SCF}$ from the change in surface energy when a charge $\sigma = e\Sigma$. per unit area is induced on the metal surface by an external field created by a charged plane located far away from the metal surface.

Equations (19) and (21) are formally equivalent, so that $dU_s/d\sigma\,|_{\sigma=0} = X^m - \bar{\mu}_e$. This relation can be generalized for an arbitrary surface charge, with X^m being simply replaced by $X^m(\sigma)$. Since only $X^m(\sigma)$ may change with charge, this relation provides an alternative to Eq. (20) for its calculation as a function of charge. In particular, one can obtain in this way the quantities $Z_0(\sigma)$ and $X_0(\sigma)$ defined in Eqs. (10a) and (10b) by using a relation derived from Eq. (21), as shown recently by Russier and Rosinberg. The values of Z_0 and X_0 obtained by following this route are very accurate, even if Smith's method is used.

There exists also an important relation between the bulk energy per electron and the potential drop between the bulk of the metal and the jellium edge, called the Budd–Vannimenus sum rule. These points will be further developed in the discussion of the results for the metal/solution interface.

If Eq. (19) were to be discussed in connection with experiments at the metal/vacuum interface, one faces again the problem that experiment gives information only on the global quantity Φ. Then, together with reasonable predictions for Φ, internal coherence of the model should be an indication that X^m is itself not too badly calculated.

For the purpose of discussing the interfacial capacitance, the relevant quantity is $X^m(\sigma)$, which represents the generalization of X^m for a charged interface, as discussed in Section 11.2 (Eq. 7a). However, $X^m(\sigma)$, as well as X^m, cannot be determined directly from experiment. However, by using a given model, they can be calculated from Eq. (6) or the generalization of Eq. (21). Then one may compare the theoretical results for $\delta X^m(0)$ to electrochemical estimates such as those made in Ref. 6. A more direct (and more demanding) comparison to experiment is the prediction of effects related to the crystallographic orientation of the surface of the electrode. For example, the predicted order of magnitude of the variations of $\delta X^m(0)$ with the exposed face may be compared with the observed change in the potential of zero charge.

(iv) Discreteness of Ions

(a) Effect of the ionic lattice

In the jellium model, the nature of the metal appears only through the value of the electronic bulk density n. While such a picture may be convenient at least as a first approximation for the surface energy and work functions of the alkalies, for which the discrete nature of the ions can be neglected, lattice effects must be reintroduced for a reasonable description of the surface properties of most metals used as electrodes in electrochemical cells.

In order to take into account the discreteness of the ionic cores, the electron–ion interaction at the surface is usually described in terms of *pseudo–potentials*, just as for bulk properties (the widely used "empty core model" potential introduced by Ashcroft is shown in Fig. 5). In the treatment of Lang and Kohn, the one–dimensional character of the jellium model is preserved. These authors treated the effect of the lattice by first–order perturbation theory. The perturbation of the jellium is taken as the difference δ $V_{ps}(r)$ between the pseudo–potential of the lattice and the potential due to the jellium. The first–order correction to the energy is given by

$$\delta E_{ps} = \int n(z) \delta V_{ps}(z)\, dz \qquad (22)$$

In this expression, $n(z)$ is the zero–order electronic density profile determined for the jellium, and $\delta V_{ps}(z)$ is the average in the plane parallel to the surface of $\delta V_{ps}(r)$. This first–order treatment has been criticized since δV_{ps} is in general not small compared to the unperturbed term. Perdew and Monnier have shown that it is better to include this term in the expression of the functional of the energy $E[n(r)]$ given in Eq. (14). While, in principle, the problem is now unavoidably a three–dimensional one, the simpler

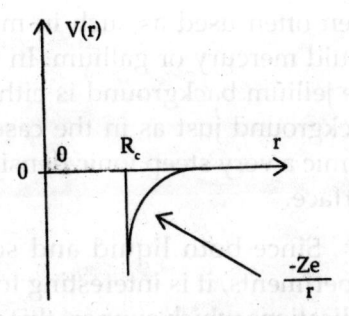

Figure 5. The one-parameter (Re) Ashcroft pseudo-potentiala. Z is the valence, and e is the electronic charge.

treatments, such as the self-consistent variational approach of Perdew and Monnier, still use one-dimensional wave functions (and hence density profiles) in Eq. (14). In the Smith-like approaches, the contribution δE_{PS} can also be included in the expression of the surface energy. Such models are sometimes referred to as "jellium-like" models. In this case, the electronic density profile differs from the jellium profile and can depend, for example, on the crystallo-graphic orientation of the surface, as in the calculation of Perdew and Monnier. This is the procedure we followed in our own work.

More sophisticated calculations have attempted to incorporate the three-dimensional nature of the semi-infinite metal. Examples include the recent work of Aers and Inglesfield or the calculation of Rose and Dobson. An alternative approach is based on the simulation of the surface by a cluster of a few atoms. Such calculations undertaken with a view to studying the metal/solution interface can be found in Refs. 83 and 84, for example. Recently, Halley and Price also used a three-dimensional version of the jellium without pseudo-potentials.

(b) Comparison between liquids and solids

By its very nature, the jellium model does not make any distinction between solids and liquids. Indeed, it has been often used as such in models for the capacitance of liquid mercury or gallium. In this view, the step profile of the jellium background is either taken as the neutralizing background just as in the case of solids or is assumed to mimic a very steep ionic density profile at the liquid metal surface.

Since both liquid and solid electrodes are used in experiments, it is interesting to ascertain whether there are indications which support this viewpoint or whether, on the contrary, there are indeed some important differences. This can be done, at least on a qualitative level, on the basis of the classical work on solid electrodes and some more recent results on liquid metal surfaces. In Fig. 6, we have plotted the results of Perdew and Monnier for the electronic density profile $n(z)$ of the (110) face (denser face in the bcc structure) of cesium. The electronic density profile $n(z)$ and ionic density profile $\rho(z)$ from the work of Hasegawa and a computer simulation by Harris *et al.* for liquid Cs are also shown.

A first observation is that $\rho(z)$ bears little resemblance to the step profile of the jellium and exhibits pronounced oscillations (this feature has been recently included in a model of the metal/solution interface by Goodisman). A more important fact in our opinion is that while for solids $n(z)$ has non-negligible values far away from the last ionic plane, $\rho(z)$ and $n(z)$ have comparable decay lengths in the case of a liquid. This may be important when the metal comes in contact with the solution. While for solids one may neglect as a first approximation the short-range interaction of the metallic ions with the solution compared to that of the electron cloud, both should be considered in the case of liquids.

(v) Extension of the Jellium Model to the Noble Metals

Silver and gold electrodes are now commonly used in electrochemical cells. Since the electronic properties of noble metals are dominated by the hybridization of states in the s band and in the d band, the formalism developed for *sp* metals cannot be used directly in this case. This issue was raised by Lang and Kohn in their seminal paper on the work function, where they found that the values of Φ obtained from the formalism for sp metals were in error by about 0.5 eV (even 1 eV with more recent experimental determinations of Φ). The theory of bulk properties of *sp* metals has been extended to the case of noble metals mainly

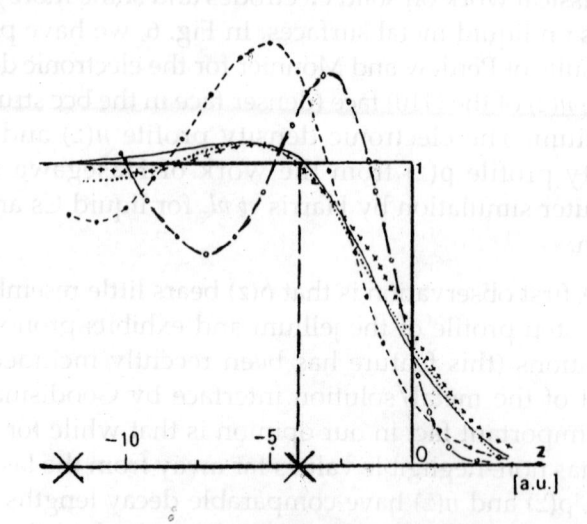

Figure 6. Normalized ionic [ρ(z)/ρ] and electronic [n(z)/n] density profiles for cesium: +++, n(z)/n for Cs(110) from Monnier and Perdew; ——, n(z)/n, and --,--, ρ(z)/ρ for liquid Cs from Hasegawa; ..., n(z)/n, and —०—, ρ(z)/ρ for liquid Cs from Harris et al. Crosses on the z axis indicate the position of ionic planes in the case of solid Cs. z = 0 is the position of the corresponding jellium background.

by Moriarty, who gave a synthesis of his work in Ref. 91, and by Wills and Harrison. An essential feature of the theory is that the effective valence of noble metals differs from the nominal valence, as a result of the s–d hybridization. For the study of surface properties of Ag and Cu, Russier and Badiali recently proposed an extension of the jellium model. The work function of Ag and Cu can be determined accurately by considering an effective simple metal contribution Φ_{ms}, which is the dominant contribution, and a potential drop $\Delta\phi_d$ due to the d–band states located on the last ionic sites of the semi–infinite lattice. In agreement with the *ab initio* calculation of Moriarty, the simple metal part is calculated by attributing to the metal an effective valence Z_s=1.5 instead of the nominal valence

$Z = 1$. An approximate version of the tight binding theory based on a sum rule due to Friedel can be used in order to compute $\Delta\phi_d$. Since the latter is much smaller than the potential drop arising from the simple metal part, one can neglect its variation with the charge at the metal/solution interface, as a first approximation.

Some Results for the Metal/Vacuum Interface

(i) Work Functions

We will not give here a detailed account of the experimental situation. Table I gives the work functions of four selected metals, including liquids (Hg, Ga) and solids (Ag, Al). The first three metals are those typically used in electrochemical experiments. Results for Al are included, since this polyvalent metal with a high electron density ($r_s = 2$) is often regarded as a good candidate for testing models. The values for solids are for monocrystalline surfaces.

Table 1
Experimental Work Functions as a Function of the Bulk Electron Density n

Metal	$10^3 n$ (a.u.$^{-3}$)	Φ^{exp} (eV)
Hg	12.7	4.48–4.52
Ga	22.3	4.30
Ag(111)	8.8	4.46
Al(111)	26.9	4.24, 4.26

For the metals in Table 1 the experimental values for the work function differ by less than 0.3 eV, and some of them even by less than 0.1 eV. For solids, an interesting indication is the sequence of values of Φ for different surface orientations. Some experimental data for Ag and Al are shown in Table 2. For Ag, the spread of the experimental values is narrow. For Al, two sets of values, labeled b and c, may be distinguished. In values in footnote c, the work functions are ordered in the sequence Φ (111) > Φ (100) > Φ (110) as observed for Ag while in values in footnote b this sequence is not obeyed. This discrepancy has been analyzed. Table 2 suggests in turn that the magnitude of the variation of Φ with the crystallographic orientation is at most 0.3 eV.

Table 2

Effect of Crystallographic Orientation on Φ

Face	Φ^{exp} (eV) Ag	Al
111	4.46	4.24, 4.26
100	4.22	4.41, 4.20
110	4.14	4.28, 4.06

The information contained in Tables 1 and 2 provides some guidelines for selecting theories. In principle, the accuracy of calculated work functions should be better than 0.1 eV in order to describe correctly the variations in Φ with the nature of the metal. This can hardly be fulfilled in practice, because Φ results from the cancellation of two relatively large quantities, as discussed previously. At present, a deviation in the range 0.1–0.2 eV should be satisfactory. For solids, when the order in the values of Φ for different orientations of the surface is well established experimentally, a more qualitative test is that the values predicted for different orientations should be properly ordered.

Calculated work functions are compared to experiment in Table 3. From the values in the table, we may draw the following conclusions:

1. In the case of solids, calculated values of Φ agree fairly well with experiment. For examples, the values of

Table 3
Comparison of Theoretical and Experimental Work Functions

Metal	Work function (eV) Theoretical	Experimental
Solids		
Al(111)	4.05, 4.27	4.24, 4.26
Al(100)	4.2, 4.25	4.41, 4.20
Al (110)	3.6, 4.02	4.28, 4.06
Ag(111)	4.40	4.46
Ag(100)	4.29, 4.95, 4.2	4.22
Ag(110)	4.10	4.14
Liquids		
Al	4.52, 3.64	4.19, 4.28
Hg	3.48	4.50
Ga	4.14	4.30

Monnier *et al.* agree within less than one-tenth of an electron volt with the experimental values in footnote *d* of Table 3. However, deviations as large as 0.2 eV are observed with the data in footnote *c* of Table 3. In the case of silver, the calculated values of Φ reported by Russier and Badiali are also in good agreement with experiment. However, in the case of Ag(100) the value of Aers and Inglesfield obtained by a three-dimensional calculation is much larger than those of Arlinghaus *et al* (three-dimensional slab geometry) and Russier and Badiali (one-dimensional jellium-like metal). We mention that the first two authors quote a value V^{exp} = 4.64 eV from, much higher than that given in the table [see also the discussion of Valette for $\Phi^{exp}(Ag)$].

Concerning the effect of the surface crystallographic orientation, one can see that for Ag and Al (footnote *d* in Table 3), predicted values of Φ for the three low-index faces are correctly ordered, following the sequence Φ (111) > Φ (100) > Φ (110).

2. For liquids, due to an insufficient understanding of their surface structure, only rough estimates may be given. From the values in Table 3, it is clear that the situation is much less satisfactory than in the case of solids. As an example, a large spread of values can be observed for Al. More generally, predicted work functions differ from experiment for all metals by several tenths of an electron volt.

From these comparisons, one may conclude that theory appears to predict correct values of the work function for solids. However, it should be borne in mind that what is ultimately needed at the metal/solution interface is a good description of the variation with charge of $X^m(\sigma)$, which at the pzc is only a part of the work function. At this point, we may recall that similar values of Φ can be obtained with quite dissimilar values of X^m, largely because of the cancellations in (D that we discussed previously. The only test at present is the comparison of different theoretical determinations of $X_0(\sigma)$, defined in Eqs. (10).

(ii) Charged Surface

At the uncharged surface, the position $X_0(0)$ of the image plane was obtained for the first time in the jellium model by Lang and Kohn. They defined $X_0(0)$ as the center of mass of the infinitesimal charge $\delta n(z)$ induced by a weak uniform field, $X_0(\sigma) = \int z\delta n(z)\, dz / \int \delta n(z)\, dz$, this definition being equivalent to Eq. (10d). For typical metallic densities, Lang and Kohn found that $X_0(0)$ depends slightly on the average free electron density and ranges from 1.6 a.u. (Al, $r_s = 2$) to 1.2 a.u. (Cs, $r_s = 6$). These values of $X_0(0)$ are measured from the jellium edge, the latter being at half an interplanar spacing from the last ionic plane.

Since the work of Lang and Kohn, the self-consistent calculation of $n(z)$ and $X_0(0)$ has been extended to the case of a charged surface by several authors (see the work of Gies and Gerhardt and of Schreier and Rebentrost, for example). These calculations may be used to test results obtained by approximate calculations as in the variational Smith–like approaches. In this case, $X^m(\sigma)$ calculated directly in Eq. (20) with trial density profiles may be inaccurate. An alternative route is to take advantage of the so–called Budd–Vannimenus sum rule, which expresses part of the potential drop in terms of properties of the bulk metal. At the charged metal surface, this relation, sometimes referred to as the "half–moment condition," can be written as

$$\phi(0) - \phi(-\infty) = -2\pi\sigma^2/n - n\, d\varepsilon(n)/dn \qquad (23)$$

where the term on the left–hand side is the potential drop between the jellium edge and the bulk of the metal, and ε is the energy per electron in the homogeneous metal. This condition allows $Z_0(\sigma)$ to be calculated from the alternative expression

$$Z_0(\sigma) = -\sigma/2n + 1/\sigma \int_0^\infty z[n_{\sigma=0}(z) - n_\sigma(z)]dz \qquad (24)$$

This route has been followed by several authors. The reader is referred to the recent review by Kornyshev for details and for a discussion of the proper use of this sum rule. Without invoking the Budd–Vannimenus sum rule, Russier and Rosinberg calculated $Z_0(\sigma)$ in the variational Smith scheme and compared their results to those obtained by Gies and Gerhardt, Who used an LK scheme. By using a special version of the Δ_{SCF} route for Φ, they have derived an expression of $Z_0(\sigma)$ which exactly reduces to Eq. (24). Their results compare well with the self–consistent calculations. As an example, we have plotted in Fig. 7 $Z_0(\sigma)$ and $X_0(\sigma)$ deduced from the results of Refs. 34 and 78 as a function of charge for $r_s = 3$. In general, $Z_0(\sigma)$ and $X_0(\sigma)$ are decreasing functions of the charge on the electrode. The

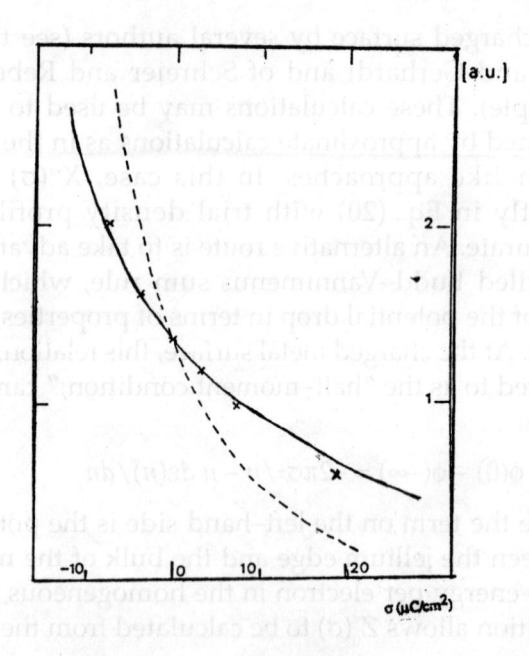

Figure 7. Center of mass $Z_0(\sigma)$ of a finite charge (——) and center of mass $X_0(\sigma)$ of an infinitesimal induced charge (- - -) for $r_s = 3$ calculated by Russier and Rosinberg. Crosses show the results of Gies and Gerhardt for $Z_0(\sigma)$.

larger increase at negative charging is generally attributed to the fact that it is easier to pull the electrons out of the metal than to push them back. Figure 8 is an illustration of the general behavior of $Z_0(\sigma)$ and $X_0(\sigma)$ for free–electron–like metals.

It can then be concluded that the weakness of the Smith–like approach can be corrected by using one of the previous alternative routes for evaluating $X_0(\sigma)$. This is important for applications to the metal electrolyte interface, since this method is much simpler than solving the self–consistent Kohn–Sham equations.

To our knowledge, no self–consistent calculations of $Z_0(\sigma)$ and $X_0(\sigma)$ of the Kohn-Sham type are available for a

Electronic Properties of the Metal Surface

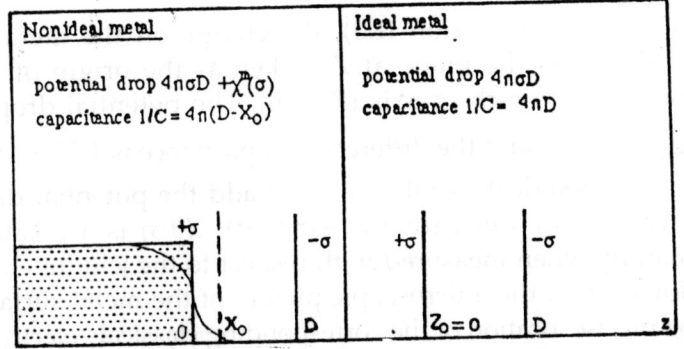

Nonideal metal	Ideal metal
potential drop $4\pi\sigma D + \chi(\sigma)$ capacitance $1/C = 4\pi(D - X_0)$	potential drop $4\pi\sigma D$ capacitance $1/C = 4\pi D$

Figure 8. Schematic view of a model charged interface. Left panel: Microscopic view of the metal; right panel: representation of the metal surface by a charged plane located at Z_0. In both panels, D is the position of an ideal charged plane.

three-dimensional solid with a surface. A three-dimensional calculation based on a different formalism was performed by Aers and Inglesfield," who calculated $Z_0(\sigma)$ for Ag(100). They suggested that the jellium calculation overestimates the variation of $Z_0(\sigma)$ by a factor of three on the basis of a comparison of their results to those of Gies and Gerhardt for $r_s = 3$ (silver with nominal valence $Z = 1$). We have checked that using the effective valence $Z = 1.5$ (higher electron density) and an electron–ion pseudopotential indeed reduces the slope of $Z_0(\sigma)$, in agreement with the suggestion of Aers and Inglesfield.

Conclusion. Relevance to Electrochemistry

In order to illustrate how the ideas presented above will be useful at the metal/solution interface, some examples are considered below.

(i) Capacitance of a Planar Capacitor

Consider the charged interface represented in Fig. 8. An ideal charged plane with a charge $-\sigma$ per unit area is located at a distance D from the metal bearing an opposite charge σ. In the classical picture of the metal (see the discussion in Section II.3), the charge on the metal is localized on the plane at Z_0, taken as the origin of the coordinates on the z axis ($Z_0 = 0$). The potential drop is $\Delta\phi_s^m = 4\pi\sigma D$, and the differential capacitance is $1/C = 4\pi D$. For the nonideal metal, one must add the potential drop $X^m(\sigma)$, whose derivative is $-4\pi X_0(\sigma)$. $X_0(\sigma)$ is a positive quantity when measured with respect to the jellium edge. However, in this microscopic picture of the metal surface, the precise location of the corresponding classical "electrode plane" is not well defined relative to the surface of the metal, represented by the position of the jellium edge or that of the last crystallographic plane. Any reasonable choice for this classical value of Z_0 would locate it between the last ionic plane and the jellium edge. Thus, the classical value of $1/C$ will always be reduced by the presence of $X^m(\sigma)$. For example, if both D and X_0 are measured with respect to $Z_0 = 0$ taken as the jellium edge, the capacitance at the pzc is $1/C = 4\pi[D - X_0(0)]$. For a macroscopic capacitor, D is many orders of magnitude larger than X_0 so that the correction is truly negligible. The situation is now very different at the interface. There, one may imagine D as being the position Z_{oHp} of the outer Helmholtz plane. Since $X_0(\sigma)$ can be an appreciable fraction of Z_{oHp}, we expect this metal nonideality to have a substantial effect on the capacitance of the interface. Provided that $X_0(\sigma)$ is not dramatically reduced by the presence of the solution, the relaxation of the surface dipole of the metal with charge should then increase the inner–layer capacitance compared to that predicted by using the traditional picture of the electrode.

(ii) Predicted Slope of the Capacitance

Insofar as the variation of $X_0(\sigma)$ with charge will not be qualitatively changed by the presence of the solution compared to the metal–vacuum case, one may infer from Fig. 7 the resulting effect on the slope of the inner–layer capacitance. Quite generally,, we may write $1/C_i = 1/\tilde{C}_i - 4\pi X_0(\sigma)$, where $\tilde{C}_i$ contains any other contribution to the capacitance (stemming from the effect of the interfacial field on the solvent polarization or the excess charge in the solution). The simplest way to picture the effect of $X_0(\sigma)$ is first to consider that $\tilde{C}_i$ is constant. In this case, the resulting C_i is a decreasing function of charge (Fig. 9). From this picture, one may infer that deviations of the experimental slopes from this behavior may be attributed to nonlinear effects in $P(z)$ or a possible variation with charge of the center of mass Z_s of the excess charge in the solution, as discussed by Kornyshev and Vorotyntsev. This point will be discussed in detail in the last section of this chapter.

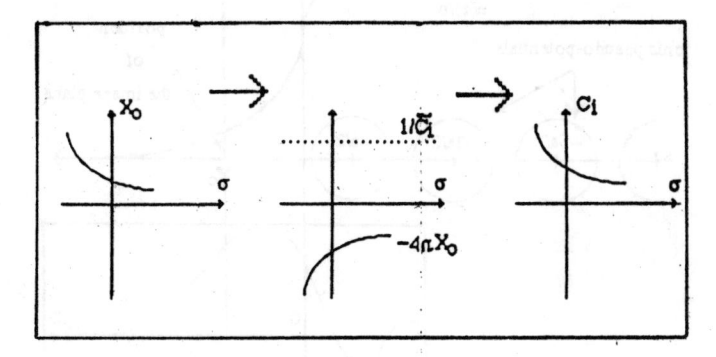

Figure 9. Predicted slopes of the capacitance C_i near the pzc assuming that (I) the metal contributes only via $X_0(\sigma)$, and (ii) all ohe rcontributions are constant.

(iii) Physical Picture of the Electrode Surface

We may now draw the picture of the metal surface (Fig. 10) which emerges from the preceding discussion. The edge of the positive background of the jellium model, $z = 0$, is at half an interplanar spacing from the last ionic plane. The position $X_0(\sigma)$ of the image plane is then located in front of this jellium edge and is directly related to the spread of the induced charge. From $X_0(\sigma)$ one obtains, for example, the asymptotic expression of the image potential on a point charge far outside the metal as $U^{im}=1/|Z - X_0(\sigma)|$, where both Z and $X_0(\sigma)$ are measured with respect to the jellium. edge. A similar dependence on $Z - X_0(\sigma)$ appears in the expression of the image potential of a classical point dipole and also in the asymptotic expression of the van der Waals forces [in the latter case, $X_0(\sigma)$ is replaced by the position of the dynamic image plane]. This picture of the metal surface may be particularly useful in the presence of adsorbed species.

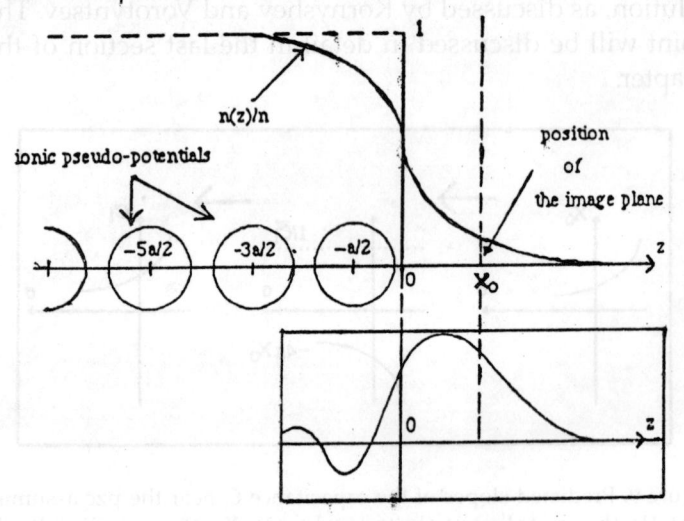

Figure 10. Schematic view of the metal, including ionic structure, with a the interplanar spacing in the direction normal to the surface. The inset shows the qualitative shape of the induced charge $\delta n(z)$.

3

Metal–adsorbate Interactions

The present section is motivated by the fact that the interaction of the metal with the adjacent liquid is an important ingredient in models of the interface. In the traditional approach, the characteristics of this interaction are estimated on the basis of electrochemical data. As mentioned in Section 1, the latter generally involve contributions from both sides of the interface. When available, independent studies which may provide more direct information on this interaction are of the greatest interest. Among such possible sources of information, some recent data from ultra–high–vacuum (UHV) experiments will first be described. Then some first–principle calculations and some related models for the metal–solvent interaction will be discussed.

Adsorption from the Gas Phase

Among the very many techniques developed recently in the field of surface science, some may be applied to the study of the electrochemical interface. For example, adsorption experiments under UHV conditions such as thermal desorption spectroscopy (TDS) or work function change ($\Delta\Phi$) measurements may provide valuable information. A general review of water adsorption at metal surfaces has been presented recently by Thiel and Madey.

TDS provides information on the metal–adsorbate bond strength. These experiments have been performed on various metals such as Ni, Rh, Ru, Ir, Pt, Cu, and Ag. For these metals, the adsorption energy is estimated to be between 48 and 57 KJ mol–1. For all metals studied in UHV, the adsorption energies are in the same range and correspond to the upper part of the physisorption range. Note that, as noted by Trasatti, this is in line with the estimated variation of the adsorption energy by a few kilojoules per mole from Hg to Ga. It is also interesting to note that silver seems quite distinctive. For this metal, only one peak attributed to desorption of water from ice layers is observed in the desorption spectrum. This feature is interpreted as indicating that for silver, water–metal interactions are, at most, as strong as water–water interactions.

Measurements of work function change upon adsorption represent the second group of UHV experiments that we would like to mention. The observed initial decrease of the work function is usually related to the potential drop due to the orientation of the physisorbed molecules. The molecular dipole points toward the vacuum; that is, on average the oxygen atom is in contact with the metal. This interpretation requires some assumption about charge transfer and neglects any contribution from a charge redistribution in the metal, similar to that giving rise to $\delta X^m(0)$ as discussed in Section II.2. The final saturation after about one layer of coverage indicates that the orienting effect induced by the surface is short-range. Finally, it is interesting to note that in the case of silver, A(D curves indicate that the orientation of water is less pronounced than in the case of copper. This seems to confirm a weaker water–silver interaction.

Discussion

The strength of the metal–solvent interaction is a question of central importance in the classical analysis of electrochemical data. It was recalled in Section I that information on this strength was obtained to some extent from the analysis of the components of the electrode potential, that is, the term $X^m(\sigma = 0) - g_{dip}^s = X^m + \delta X^m(0) - g_{dip}^s$. From estimates of $\delta X^m(\sigma = 0) - g_{dip}^s$, the metals have been classified on a "hydrophilicity scale. By assuming that the variation of this quantity with the nature of the metal is mainly due to a metal–induced orientation of the solvent dipoles through g_{dip}^s, it was inferred, for example, that water is much more oriented and thus more tightly bound on Ga than on Hg.6 Thus, Ga is considered to be a hydrophilic metal whereas Hg should be hydrophobic. However, UHV experiments indicate, in agreement with other electrochemical estimates,' a relatively minor influence of the nature of these two metals on the water–metal bond strength. These seemingly conflicting findings were reconciled by considering that major variations of potential drop from metal to metal might be induced by relatively minor energy changes. This point favored the interpretation of "hydrophilicity" in a relative sense.

Since these conclusions follow to a large extent from the assumption that δX^m is nearly metal–independent, this point has been investigated by using a nontraditional description of the electrode. It is now known from theoretical calculations that $\delta X^m(0)$ is metal-dependent. Calculations indicate that $\delta X^m(0)$ may vary from metal to metal by a few tenths of an electron volt. Since this is of the same order as the estimated variation of $\delta X^m(0) - g_{dip}^s$ for different metals, the presence of this metal contribution makes the interpretation of experimental data in terms of hydrophilicity only tentative, as suggested in the early work of Badiali *et al*. At the present stage of development of theoretical calculations, it is not possible to give values of $\delta X^m(0)$ which are accurate enough to allow definitive conclusions on this point. One of the reasons for this situation is our insufficient understanding of the metal–solvent coupling. However, recent work in this field. Suggests that a new concept be introduced in the analysis of the interface: *the effect of the metal–solvent bond length.*

Quite generally, substrate–adsorbate interaction may be characterized both by the *bond strength and by the bond length* (see Refs. 116 and 117). For a single adsorbed particle *at rest*, the two quantities are indeed correlated since the equilibrium surface–particle distance corresponds to the minimum of the potential energy curve. In the case of a chemisorbed molecule, the potential energy curve is expected to exhibit a pronounced minimum and rise rapidly with increasing distance between the surface and the molecule. Then the equilibrium position of the molecule at the surface is well defined and will be only slightly affected by, say, an external field or the presence of coadsorbed particles. This is just the view which underlies the classical analysis: it is implicit in this view that the water molecule has its center of mass rigidly fixed to the metal and only its orientation may change with the nature of the metal or under the effect of the interfacial field. Therefore, while "hydrophilicity" refers to the strength of the bond, in the underlying analysis the metal affects in fact only the orientation degrees of freedom of the molecule via g_{dip}^s.

While this picture might be valid for a truly chemisorbed solvent, it need not be true for water, which appears to be not so strongly bound on the metals analyzed in UHV, especially on silver. In the case of a weakly bound

Metal-adsorbate Interactions

molecule, the potential energy curve is expected to have a less steep variation with the distance between the surface and the molecule than in the case of chemisorption. In a dense fluid, the average minimum metal–solvent distance will differ from the equilibrium value for the isolated particle at rest, since, in addition to the effect of other molecules or ions, the molecules in the liquid may cross an energy barrier of the order of their mean kinetic energy. Therefore, the average metal–solvent separation is expected to be associated with the repulsive part of the potential. The presence of the solvent at this average distance from the metal surface will give rise to the perturbation δX^m of the surface dipole. The variation of δX^m with the nature of the metal will thus be related in some ways to the variation of this metal–solvent separation, mainly through the repulsive part of the potential. Therefore, the precise value of the energy minimum for an isolated molecule will be much less relevant than in the case of a strongly bound solvent.

Theoretical Description of Metal–Adsorbate Interactions

The interaction of metal surfaces with various adsorbates has been studied from first principles both in chemisorption and physisorption situations. This aspect of the theory is a wide field of research in itself, and it is not the purpose of this section to present the state of the art in this field. Important aspects have been reviewed. Here we will present briefly some studies on water adsorption on metal clusters and a typical physisorption calculation.

(i) Cluster Calculations

Theoretical studies of water adsorption on metal clusters are rather scarce (see the review by Muller). This kind of calculation has been performed for one water molecule on clusters of Pt, Al, and CU. See also the recent work of Kuznetsov et al. who also considered Hg, Cd, Zn, and Au along with the case of water clusters. The methods of quantum chemistry are used in order to study the interaction of the water molecule with a cluster of a few metallic atoms simulating the immediate environment of water adsorbed at the metal surface. From these calculations, the following findings emerge: (1) water is adsorbed without dissociation or significant deformation of' the molecule and with the oxygen atom toward the metal; (2) the binding involves mixing of the water lone pairs with molecular orbitals of the cluster; (3) the interaction potential involves a repulsion between metal electrons and the closed-shell orbitals of' the molecule, similarly to the case of a rare gas; and (4) predicted binding energies are on the order of those estimated in UHV but depend on the crystallographic definition of the adsorption site.

As is well known, these calculations raise the question of the relevance of results obtained with a cluster containing a small number of atoms to the case of an extended surface. The latter situation involves effects related to delocalized electrons and to long-range electrostatic forces which may be absent in cluster calculations. For example, the total dipole moment of the water–cluster system leads to work function changes much greater than those estimated from UHV. Another aspect specific to the metal/solution interface is that we are in fact interested in the effective interaction of a given molecule in the presence of the other molecules of the liquid. Since the molecule does not form a true chemical bond with the metal, one may expect that the presence of this dense phase will first modify the distribution of the free electrons on the metal surface so that the effective interaction will also reflect this effect. These considerations are clearly beyond the cluster approach. Then what one may do is to try to incorporate some results of these calculations in a model for the interaction at the metal/solution interface. The qualitative indications concerning the repulsion can be included in models of the potential in a rather simple way. In contrast, the attractive contribution is clearly not amenable to simple modeling. However, if one recalls the discussion given earlier on the equilibrium metal–solvent distance, the details of the attractive contribution may be not so important for a weakly bound solvent.

(ii) Metal–Rare–Gas Interaction

The interaction of a rare–gas atom with a metal surface has been studied from first principles by Zaremba and Kohn. They have shown that, to a good approximation, the interaction potential $V(Z)$ of an atom located at a distance Z from the metal surface can be written as

$$V(Z) = V_{HF}(Z) + V_{corr}(Z) \tag{25}$$

The first term, $V_{HF}(Z)$, is essentially a short–range repulsion due to quantum exchange between the conduction electrons of the metal and the closed–shell orbitals of the rare gas. It corresponds to a treatment in the Hartree–Fock approximation. Harris and Liebsch have extended the treatment of Zaremba and Kohn in order to include correlations between the metal electrons. They introduced a pseudopotential V^{PS} which they treated by perturbation theory. They have shown that a first approximation is to consider V^{PS} as a local operator. Then, instead of $V_{HF}(Z)$, they obtain a repulsive term of the form

$$V_R(Z) = \int n(\mathbf{r})\, V^{PS}(\mathbf{r}, \mathbf{R})\, d\mathbf{r} \tag{26}$$

where $n(\mathbf{r})$ is the electronic density profile of the metal in vacuum, and $\mathbf{R}$ is the coordinate of the atom. Note that they mention that this local approximation may not be sufficient if one seeks the detailed description of $V(Z)$ required for interpreting the helium diffraction experiments for which the theory was initially devised In our case, however, we may still use it for qualitative purposes.

In Eq. (25), the second term, $V_{corr}(Z)$, is an attractive contribution related to long-range electronic correlations between the metal and the atom. It corresponds to the presence of dispersion forces as in the van der Waals potential. Zaremba and Kohn have shown that the calculation of $V_{corr}(Z)$ requires the knowledge of the dynamic response function of the metal and the atomic polarizability. An expansion of the general expression of $V_{corr}(Z)$ in inverse powers of Z gives as the leading term the van der Waals energy.

$$U_{VdW}(Z) = C_3 / \left|Z - Z_0^{dyn}\right|^3 \tag{27}$$

where the distance Z is measured with respect to the dynamic image plane position Z_0^{dyn}, which, as well as the strength C_3, can be calculated from the theory. The approach of Zaremba and Kohn presents two difficulties: on the one hand, the evaluation of C_3 and Z_0^{dyn} for a real solid is not straightforward. On the other hand, $U_{VdW}(Z)$ exhibits a nonphysical (artifactual) divergence for $Z \to Z_0^{dyn}$. Annett and Echenique have derived an alternative expression which avoids these difficulties. The interested reader is referred to their paper for details. The important steps in the calculation are given in the appendix of this chapter. The input parameters for the metal are the surface plasmon frequency $\omega_s = (3/2r_s^2)^{1/2}$ and the static image plane position X_0, which enter the surface plasmon dispersion relation $\omega_s(q)$. For the molecule we need the dipolar oscillator strengths f_{n0} and the related frequencies ω_{n0}. In the formulation of Annett and Echenique, U_{VdW} remains finite at all metal-atom separations while being of course equivalent at large Z to the result of Eq. (27).

Finally, it should be mentioned that while the approach of Zaremba and Kohn considers the effect of bound electrons in the metal, a straightforward application of the formalism of Annett and Echenique is possible only in the jellium model. However, a possible heuristic extension of the formalism simulating the effect of the solid is to introduce the value of X_0 calculated for the solid and an appropriate value for ω_s, through rs.

(iii) Effective Medium Theory

An alternative approach to the theory of metal–adsorbate interactions is based on the use of the effective medium theory (EMT) (see the papers of Lang and Norskov, for elxample). EMT is appropriate, in principle, both for chemisorption and physisorption situations. The main idea is that the adsorption energy of the adsorbate in the inhomogeneous electron gas at the surface can be determined from the corresponding energy $\Delta E(n)$ in a homogeneous host of uniform electron density n. To a good approximation, the adsorption position is such that the electron density is equal there to the density n for which $\Delta E(n)$ is minimum. EMT has been compared to first–principle calculations in the case of rare gases or simple atoms. The agreement found there supports the use of EMT, which indeed introduces a considerable simplification. To our knowledge, there is no rigorous justification for the use of EMT in a more general adsorption situation. At the metal/water interface, Goodisman found recently that it leads to largely overestimated adsorption energies. Thus, the appropriate use of EMT in this case still awaits clarification.

We close this section with the following remarks: because of the complexity of a first–principles calculation of the interaction of water with an extended surface, the modeling of this interaction at the metal/solution interface is unavoidable for practical reasons. The simplest way to do this is to take metal/ rare–gas–like interaction potentials. This is the route followed in most investigations of the metal/solution interface. This was done largely for simplicity. However, given the data (mostly from UHV) on the metal–water bond strength, it does not seem unreasonable–as a first approximation–to model the interaction potential by that of a rare gas, especially in the case of water on silver, for which it seems particularly weak. In addition, the truly important part of the potential for a relatively weakly bound solvent is the repulsive contribution, as discussed above. A final check of the adequacy of a given model potential may consist in verifying that the results for the capacitance show a weak dependence on details of this model potential. We will show later that this is ultimately the case by comparing the results obtained with weak adsorption potentials.

GENERAL STRUCTURE OF RECENT MODELS FOR THE CAPACITANCE

The models proposed recently for calculating the capacitance are naturally based on models for the metal, for the solution, and for their coupling. An important aspect in most of these models is that the three–dimensional nature of the interface is discarded, mostly for simplicity. The properties of the interface are assumed to vary only in the direction normal to the metal surface. However, the surface structure of the solid is usually taken into account approximately by the introduction of ionic pseudopotentials averaged in the direction parallel to the interface. A few exceptions are a recent work by Halley and Price, where solvent molecules distributed in a two–dimensional array interact with a three–dimensional jellium, and the work on small metal clusters, where the solution is absent. While in other contexts a one–dimensional picture may be considered as an oversimplified view of the interface, especially in the case of solids, the presently available three-dimensional calculations have not introduced features which may lead one to consider one-dimensional models as qualitatively inadequate for describing the capacitance. For example, Halley and Price found that the capacitance curves obtained with the three-dimensional calculation do not differ much from those obtained in the one-dimensional case. This behavior seems related to the fact that the capacitance, as a macroscopic quantity, smooths details of the interfacial structure in the direction parallel to the surface (see, for comparison, the definition of the capacitance in the work on small clusters). In another respect, three-dimensional calculations may be required in the future for a more quantitative discussion than the one we intend to present in this chapter.

The Metal Side

Most models describe the metal on the basis of the jellium model, in some cases including electron-ion pseudo-potentials. As discussed in Section III.2(iv), the jellium model is used in some models in an identical manner for solids and liquids. In the latter case, the step profile also mimics the ionic profile, whose exact form is still a subject of debate, as discussed previously. Using the step profile as an approximation to the ionic profile for liquids means also that the ionic pseudo-potentials should be uniformly distributed up to the jellium edge. Since the edge of the step profile is the plane of zero adsorption of the electrons, the system is globally neutral. The situation differs from that in the case of solids, where for reasons of electroneutrality the jellium edge is at half an interplanar spacing from the last ionic plane. Technically, most authors use the Smith scheme for describing the properties of the electron gas. Exceptions are the work of Goodisman and of Halley and co-workers. Concerning the type of metal, the formalism for sp metals is often used without great care for most metals, including, for example, mercury, gallium, or noble metals. In the case of mercury, even bulk properties cannot be described well within the simple metal formalism. Gallium is a metal which exhibits peculiar properties (structure factor with a shoulder, density of the solid phase lower than the density of the liquid phase, polymorphism, etc.). In order to reproduce its special behavior, one must consider the pseudo-potential as an operator in the quantum mechanical sense. 114 Finally, for noble metals, s–d hybridization is a characteristic feature [Section III.2(v)]. All these aspects are not only a matter of theoretical sophistication, especially when one seeks quantitative agreement between theory and experiment.

The Solution Side

In the first models, the solution was described in the same frame as in the classical models of the inner layer. A dielectric film of monolayer thickness was considered in Refs. 86, 98, 106, and 136. Schmickler also used a lattice model for the layer next to the electrode. Besides these monolayer models, a more general description of the solution has been considered, by extending to inhomogeneous systems the theoretical methods introduced for describing the properties of ionic solutions in bulk phase. Beyond the "primitive model" in which the solvent acts as a dielectric continuum, it is now possible to consider some "civilized models" in which the solvent is treated on a microscopic scale. The first civilized model is the treatment of the solution as a mixture of hard spheres with embedded point dipoles (solvent) or charges (ions). While this model may provide a first approximation in the case of aprotic solvents, it is known to give a poor description of bulk properties in the case of H-bonded solvents such

Metal-adsorbate Interactions

as water. The statistical mechanics of this system can be solved in the mean spherical approximation (MSA), for which some interesting properties can be obtained analytically. At the interface, the results of the MSA are restricted to small surface charges. In this case, the solution contributes to the capacitance by a constant term. In the case of an arbitrary surface charge, Schmickler and Henderson have devised a heuristic extension of the MSA.

The MSA has also been used in studying the physisorption of polar molecules on a neutral wall. The MSA predictions have been compared to Monte Carlo simulations. While the structure of the interface is poorly described, the correct order of magnitude of the potential drop is obtained. Some shortcomings of the MSA have been detailed. In particular, it has been shown that the MSA cannot reproduce some aspects related to the image potential.

In the bulk phase, the same model for the solution has also been treated with the hypernetted chain approximation (HNC). The HNC has been also used in studying some models in which a more sophisticated description of the solvent is considered. These models may include the polarizability of the molecules or some quadrupolar effects. No HNC results exist for a planar charged interface. In recent work, attempts to mimic this interface by considering a solution in contact with a very large charged sphere have been described. Recently, it has been shown that the HNC suffers from the same shortcomings concerning the image potential as the MSA. Moreover, in the case of a pure polar fluid, the HNC does not give the exact behavior of the profile far from the wall.

The Coupling at the Interface

In all the models proposed so far, each side of the interface responds only to changes in the average distribution of the particles on the other side (mean–field coupling). As discussed in Ref. 144, this is equivalent to a treatment of the electron–solution interaction by first–order perturbation theory. Such an approximation will be valid only in the case of weak interfacial coupling.

Since these models are primarily concerned with interfaces without specific ionic adsorption, they do not consider short–range interactions between the metal and the ions in the solution. The metal–ion coupling is thus purely electrostatic and adds nothing to what has been discussed in the case of a charged surface [Section III.3(ii)]. Since the electron density has very small values at the position of the ions, it is sufficient for many purposes to replace the excess charge in the solution by a charged plane.

The models of the coupling are then mainly models for the metal–solvent interaction. In Section IV.3(ii), the molecule–metal interaction in the gas phase was examined. The same ingredients are retained in the models for the metal in contact with a dense phase. The models are usually developed following the example of the physisorption potential, which involves a repulsive part and an attractive one determined separately.

(i) Repulsive Contribution

A very common feature of the models for the coupling at the interface is the incorporation in the electronic energy functional [Eq. (14)] of a repulsion of the metal electrons by the electronic cloud of the solvent molecule. Various forms of this repulsion are considered on semiphenomenological grounds. These potentials or "pseudo–potentials" are characterized by a given shape and strength. The unknown parameters are either determined by fitting to experimental data or left unknown, with their effect on the results being investigated. The simplest simulation of the presence of the solvent is a repulsive potential, for example, a square barrier. The metal electrons are placed in a potential given by $V^{sol}(z) = V_0 \theta(z - Z_{sol})$, where the parameter V_0 controls the intensity of the repulsion and where Z_{sol} is the edge of the solvent. This form was used by Schmickler and Henderson and by Halley et al. Another model is the barrier of constant slope, $V^{sol}(z) = V_0(z - Z_{sol})$, used also by Schmickler and Henderson. In this case, Z_{sol} is the jellium edge. For a summary of the work of Schmickler et al.

The discreteness of the solvent was taken into account by the introduction of electron–molecule "pseudo-potentials." Badiali et al and Feldman et al. used the Harrison point potential $V^{ps}(r, R) = \lambda \delta(r - R)$ between an electron at r and a molecule at R, where λ is a parameter controlling the intensity of the repulsion. Halley and Price constructed from first principles a neon pseudo–potential in order to simulate the interaction between the closed-shell water molecule and the metal. It contains, in principle, no adjustable parameters. Their results are qualitatively similar to those presented in Section IV.3(ii) for metal–rare gas interaction. In our recent work, we used a rectangular potential defined by $V^{ps}(R, r) = V_0$ for $|R - r| \leq R_0$ and equal to 0 otherwise, where R0 is the radius of the solvent molecule. This form can be viewed as the average of the Harrison point potential over the volume of the molecule.

(ii) Attractive Contribution

Attractive contributions are not treated explicitly in the work of Schmickler and Henderson. Attractive contributions to the interaction were considered qualitatively by Kornyshev and Vorotyntsev. Their quantitative effect was included in a simple way in the work of Feldman et al. They are also implicit in the recent work of Goodisman based on the effective medium theory.

In the work of Halley *et al.*, the attractive forces include dipole–image dipole interactions and constant force, or they are built into the pseudo–potential. By using some adjustable parameters, Halley et al fit an estimated binding energy of water on some metals of about 20 kT. In our work, we used the parameter–free van der Waals potential discussed in the previous section. For any reasonable value of the repulsion parameter V_0, the depth of the potential is much smaller. This choice should be appropriate to less tightly bound water (presumably water on silver).

(iii) Equilibrium Metal–Solvent Distance

An important point is that the equilibrium configuration of the metal–solvent system results from their mutual interaction; that is, if one considers the effect of the solvent on the metal, one should also consider the reverse effect. A very important point must be borne in mind: the metal–solvent interaction must be determined in a self-consistent way. On the one hand, a solvent molecule in the fluid feels from the metal side an average potential $V_{ms}(Z)$, which is a generalization of the metal–molecule interaction $V(Z)$ defined in Eq. (25) and pertaining to an isolated molecule. On the other hand, the electron gas responds to the Presence of the solvent, so $V_{ms}(Z)$ must be a self-consistent potential.

Although $V_{ms}(Z)$ includes some effects of the other solvent molecules and is possibly charge–dependent, it exhibits the same kind of variation with Z as $V(Z)$. In other words, $V_{ms}(Z)$ does not behave like a pure hard wall potential; on the contrary, it acts as a "soft wall" just like $V(Z)$. Nevertheless, by using some appropriate criterion, we can associate with the soft wall $V_{ms}(Z)$ an equivalent hard wall from which it is possible to define a minimum metal–solvent distance as discussed in Section IV. 2. In our work the rigid wall is introduced just as an intermediate quantity

Metal-adsorbate Interactions

in the calculations. However, since the rigid wall, concept is commonly used, hereafter our discussion will be in terms of the position of the rigid wall d, or, equivalently, in terms of the metal–solvent distance of closest approach.

Within this representation of the interaction, it is clear that the position of the rigid wall cannot be arbitrary, given the requirement of force balance at the interface. This can be visualized by the following thought experiment. Suppose that the interaction is tuned gradually by varying the strength of the repulsion from 0 to its actual value, say, V_0. In a model where the wall is fixed, as a result of increasing repulsion from the molecules, the electronic profile in the metal will adjust until it reaches its final value $n_{v_0}(z)$. In t real physical situation, increasing V_0 will also increase the repulsion felt by the solvent molecules. Thus, they will move away from if metal surface, with a subsequent decrease of the actual repulsion between the metal and the solvent. At equilibrium, the final value of the electronic profile will certainly be different from $n_{v_0}(z)$ for the previous case. This means that considering the position of the wall as fixed with respect to the metal surface will lead to the interface being overconstrained.

Of course, attractive forces also play an important role in the determination of the equilibrium (together with the pressure of the fluid). In the case of chemisorption, the solvent molecules are tightly bound to the metal by chemical bonds, and it is reasonable to assume that the metal–solvent distance is fixed. For the interfaces considered in this chapter, however, some experimental evidence favors an interaction more characteristic of physisorption. Then the average metal–solvent distance must be considered as a quantity which must be calculated self-consistently with the other properties of the interface. Taking the boundary for the solvent as fixed to the jellium edge, as Schmickler and Henderson do, cannot be justified a *priori*. This assumption will have serious consequences at a charged interface: in this case, the variation of the metal–solvent distance with charge will eventually lead to an analysis of capacitance curves which will be qualitatively different from that for the case of a fixed metal–solvent distance.

This is the important point on which the models differ. In most of them, the particles in the solution are forced to occupy only one half–space limited by a rigid wall simulating the metal surface. A central question then is the precise position of this wall at each surface charge. If the position of the rigid wall is fixed, the only remaining problem is to determine the quantity X_0 in the presence of the solution. X_0 contains then *all the effects of the metal* (except a much smaller influence of the electron tail on the polarization). In the other case, the position of the wall and X_0 should be determined within the same model of the interface.

In practice, the coupling introduces additional terms in the metal energy functional [see Eq. (14)], including terms due to the ions (electrostatic) and to the solvent (effect of the solvent polarization and short–range repulsion). The effect of the electron spillover in the solution is important with regard to the Polarization, in contrast to the reverse effect. This point, which has been checked explicity, may be understood in terms of the energies involved. Most of the electrons in the interfacial region experience an average effective potential per electron of the order of an electron volt (~40 kT). This is much larger than the effect of the electric field due to the dipoles. In contrast, the coupling of one dipole with the tail of the electronic Profile introduces an additional energy that is comparable to kT and is hence a significant part of its mean kinetic energy.

Two different cases depending on, whether the metal–solvent distance is maintained fixed or not will now be discussed.

MODELS WITH FIXED METAL-SOLVENT DISTANCE

Shortly after the work of Badiali et al. appeared, 148 Schmickler and Henderson (SH) proposed a similar mode187 coupling the jellium description of the metal to the ion–dipole model for the solution. In addition to some technical details (discussed, for example, in Refs– 87, 67, and 150), the major difference between the two approaches lies in the Position of the hard wall for the solution. While this quantity appeared as a crucial parameter in the work of Badiali et al., 148 SH always set the Position "d" of the hard wall at the jellium edge, independently of charge. It has been mentioned previously that unless the solvent molecules are tightly bound the ionic skeleton by chemisorption, there is no physical reason which justifies this assumption. In addition, the jellium edge is located at one half interplanar spacing from the last ionic plane in the metal. In the choice of SH, it is implicit that the distance d is determined by the interaction of the solvent with the metal ions. The value of d should then be linked with some characteristic property of the ion rather than with the jellium edge. Some correlations between the interplanar spacing and the diameter of the screened ion indeed exist, but they are not strictly equal.

Schmickler and Henderson considered a possible variation of d with charge. By assuming that the wall Position is such that the electron density $n(z = d)$ (which depends on charge) at the Position d takes the value $n(z = 0)$ at the jellium edge for the uncharged surface, they claimed from the results of this model that such a variation is a second–order effect. It is not clear whether or not SH repeated the whole (self–consistent) calculation with this new assumption for d. The hypothesis that $n(z = d)$ should be equal to the density at the jelliurn edge independently of charge cannot be justified. As detailed below, any reasonable model for the metal–solvent interaction leads to a variation of d with charge much more pronounced than that estimated in the work of SH More recently, in a comment''' of the analysis made by Kim et al. of the SH approach, Schmickler and Henderson estimated that the maximum variation of d should be about 0.2 A. In fact this is in line with our results. Such a seemingly small variation can have a very significant effect on the charge–capacitance curve and on the influence of the nature of the electrode.

Since the metal–solvent distance is fixed in the SH approach, the metal contributes to the capacitance by an amount $1/C_m = -4\pi X_0$. Then the important parameter is the bulk electronic density of the metal. SH found that at the pzc, large inner–layer capacitances correlate with large electron densities. While the effect of the electron density via X_0 is indeed the expected one (see the discussion in Section III.4), the apparently satisfactory results of SH for some sp metals would mean that the distance d at the pzc is almost the same for all these metals. There is no obvious explanation for this behavior [in the work of Badiali *et al.* he sequence $C_i(Hg) < C_i(In) < C_i(Ga)$ was obtained largely from the differences in the ionic radii]. Given the large theoretical uncertainties in the description of liquid Hg and Ga, a definitive answer to this question cannot be given now. A discussion on a qualitative level will be made in the next section.

The MSA treatment of the hard–sphere ion–dipole mixture gives a constant contribution of the solution to the capacitance. Then the sole variation of X_0 is far from being sufficient for predicting the correct behavior with charge. This point was recognized by SH, who devised an extension of the MSA result mainly through a Langevin distribution for the polarization. While the curves exhibit now a pronounced hump due to the solvent, agreement with experiment is still not satisfactory. In our opinion, this is largely due to the neglect of the additional charge dependence related to the variation of the wall position with charge.

Besides studying the capacitance, Schmickler and co-workers used the jellium–ion–dipole model to investigate other interfacial properties. They found that this model, and Particularly the improved description of the electrode, proved to be useful for understanding some experimental observations.

MODEL WITH VARIABLE METAL-SOLVENT DISTANCE

Determination of the Distance of Closest Approach

The importance of the variation of d with charge has been recognized by several authors. This point was discussed on a formal level by Kornyshev et al. A very simple estimate of this effect based on the assumption of a linear variation was made by Badiali et al. This problem was discussed again qualitatively by Kornyshev and Vorotyntsev, 106 and quantitative estimates based on their ideas were made in the work of Feldman et al. These studies were later continued by Halley and co-workers. More recently, our own work has been mainly devoted to the study of this effect and its consequences for the analysis of charge–capacitance curves for various metals and solvents. These studies will not be detailed here. Only some aspects will be discussed using the example of our own work. Among others, the following two important conclusions were reached in these studies:

1. The metal–solvent distance of closest approach, d, varies with the charge σ. This conclusion follows from the force balance condition at the interface as discussed in Section V–3(iii). A good discussion of the qualitative aspects of this effect has been presented by Kornyshev and Vorotyntsev. A more systematic analysis was later made in the work of Halley *et al.* and in our work. The interfacial electric field affects the force balance in two different ways. On the one hand, the metal–solvent molecule potential will change with charge as a consequence of the variation of the electron density profile of the metal: at positive charging, the electrons are pulled back into the metal. Then a solvent molecule at a given position feels less repulsion from the electron cloud. The converse effect occurs at negative charging: a larger spillover of the electrons from the surface means that a stronger repulsion is felt by the solvent molecules. On the other hand, coulomb forces between the net charge on the metal surface and the opposite charge in the solution will result in an electrostatic pressure pel at the interface, similar to that existing between the plates of a classical condenser. At the interface, the motion of the solvent molecules is correlated to that of the ions; one may picture this by considering an ion and its solvation shell. Accordingly, the molecules will also be subject to this pressure, and thus a compression of the interfacial region may occur. Note that this effect is made possible because the metal behaves like a soft wall for the molecules. Thus this effect may exist even if the molecules are considered as rigid. Without invoking a solvation process, a quantitative account of this intuitive view of the interface based on a statistical mechanical derivation of the electrostatic pressure of an assembly of polar molecules subject to an applied field will be given below [see Eq. (28) or (30)].

2. The overall shape of the $d(\sigma)$ curve does not depend on details of the metal–solvent interaction. This second

important conclusion, made by several authors, is valid provided one is dealing with physisorption. This point results from the fact that the force balance is determined by basic physical properties of the interface. At large enough magnitude of the charge, a decrease in the metal–solvent distance is expected from the dominant effect of the electrostatic pressure Pel, which varies as 0–2. This behavior is similar to the contraction of a classical condenser upon charging. At moderate charges, a very important feature is the asymmetric behavior of $d(\sigma)$ with respect to the sign of the charge. At positive charging of the metal, a decreasing repulsion of the solvent molecules by the electron cloud enhances even further the contraction effect due to the electrostatic pressure. The value of d will always decrease from its value $d(\sigma = 0)$ at the pzc. At moderate negative charging, the electron spillover increases and so will the value of d. This behavior is the continuation of the trend on the positive side of tile pzc. However, this increase of d competes now with the decrease due to P^{el}. One thus expects the value of d to first reach a maximum when the two opposite effects exactly cancel, and later to decrease again, due to the ultimate trend imposed by P^{el} at high enough negative charges.

This behavior has been reported by several authors. Kornyshev and co-workers as well as Halley *et al.* define the metal–solvent distance d as the value at which the surface energy, including the interaction of the metal with the solution, is minimum. Such a procedure means that the total metal–solution interaction energy, defined as the difference between the energy of the metal in the presence of the solution and that in its absence, is minimum with respect to d. In our own work, a slightly different approach was followed. By using a statistical mechanical method based on the use of the Gibbs and Bogolioubov inequality, we defined a distance of closest approach of the solvent molecules to the metal surface precisely from a force balance condition:

$$P^{ref} + 2\pi\sigma^2 + \int_d^\infty dZ \rho(Z-d) \, \partial V_{ms}(Z)/\partial Z = 0 \qquad (28)$$

where P^{ref} is the bulk pressure of the fluid, $\rho(Z-d)$ is the density profile for solvent molecules at the charged interface, and $V_{ms}(Z)$ is the metal–solvent potential. In15deriving Eq. (28), we made use of the so-called contact theorem. This theorem tells us that

$$\sum_\alpha kT\rho_\alpha(z=d_\alpha) = P^{ref} + 2\pi\sigma^2 \qquad (29)$$

where T is the temperature, and ρ_α is the value of the density profiles ($\alpha \equiv$ ion or dipole) at the wall position d_α. The left-hand side of Eq. (29) is simply the momentum transfer to the wall. Since $V_{ms}(Z)$ varies sharply compared to $\rho(Z-d)$, Eq. (28) suggests that we need essentially values of $\rho(Z-d)$ near $Z = d$. Then from Eq. (29) we may replace $\rho(Z-d)$ by the corresponding profile at zero charge, $\rho_o(Z-d)$, plus a correction term which is of the order σ^2. Hence, we can write Eq. (28) as

$$P^{ref} + 2\pi\sigma^2/\varepsilon_{eff} + \int_d^\infty dZ\rho_0(Z-d) \, \partial V_{ms}(Z)/\partial Z = 0 \qquad (30)$$

where the parameter ε_{eff} takes into account this first deviation of $\rho(Z-d)$ from the value at zero charge $\rho_0(Z-d)$.

Equation (28) or (30) shows that the position d of the wall is such that the total force per unit area exerted by the potential $V_{ms}(Z)$ on the fluid is balanced by the total pressure of the fluid ($P^{ref} + P^{el}$), where $P^{el} = 2\pi\sigma^2$ if Eq. (28) is used or $P^{el} = 2\pi\sigma^2/\varepsilon_{eff}$ otherwise. These equations justify the intuitive introduction of Pel discussed above.

Our definition of the distance of closest approach has a simpler physical interpretation than that defined from the minimum of the total surface energy. The distance d is now obtained from a force balance equation directly related to the metal–solvent molecule potential $V_{ms}(Z)$. The physical meaning of d can be better visualized if we note that the value of d is such that $V_{ms}(d)$ is roughly equal to kT [if the repulsive part of $V_{ms}(Z)$ is sufficiently steep, the precise value of the pressure term is not crucial for the value of d]. $V_{ms}(d)$ is an estimate of the energy barrier that a particle of the fluid may climb, and d then has the meaning of an average turning point for the solvent molecules.

As in Eq. (25), $V_{ms}(Z)$ is the sum of the dispersion term U_{vdW} and a repulsive term given by an expression similar to Eq. (26) for V_R:

$$V_{ms}(Z) = U_{vdW}(Z) + \int n(z)\, V^{ps}(\mathbf{r}, \mathbf{R})\, d\mathbf{r} \qquad (31)$$

However, $n(z)$ is now calculated for the metal in the presence of the solution. This implies that the calculation of d and $n(z)$ must be performed self-consistently since the electronic profile is now determined from a functional of the energy $E[n(z)]$ which contains a coupling term such as

$$U^{sol}[n(z)] = \int n(z)\, V^{ps}(\mathbf{r}, \mathbf{R}) \rho(Z - d)\, d\mathbf{r}\, dZ \qquad (32)$$

At each surface charge, we can calculate $d(\sigma)$ and $X_0(\sigma)$ and later evaluate the contribution of the metal to the capacitance. Some results for the metal–solvent distance will now be presented.

The variation of d with charge has been studied quantitatively by Feldman *et al.*, by Halley and co-workers, and by ourselves. Typical $d(\sigma)$ curves are given in Fig. 11. A comparison of the curves on a quantitative level is useless since they do not correspond exactly to the same physical quantity. They have also been obtained for different metals and with different models for the metal and for the

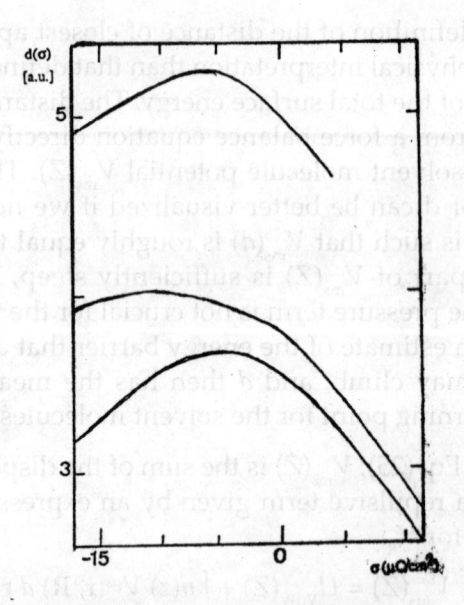

Figure 11. Results for the variation of the distance of closest approach with charge. From top to bottom: Results of Feldman et al. (r_s = 2.7), of Halley and Price (r_s = 2.59), and from our work on silver (r_s = 2.64). A solvent radius (R_0 = 1.5Å) has been added to the results of Feldman et al. In order to have the same definition of d for the three curves.

coupling. However, they have very similar shapes. In the calculation of Halley and Price, the final decrease of d on the negative side of the pzc seems to be shifted to more negative charges. A possible explanation of this feature is that their metal–solvent potential has a very steep variation with distance at negative charges. In our work, the magnitude of d depends on the strength V_0 of the metal–solvent repulsion. However, we have checked that the variation of d with charge is virtually independent of V_0.

Beyond these remarks, the observation that within different models the minimum metal–solvent distance exhibits a common behavior with charge is strong support for the previous analysis of the underlying physical mechanism. This observation is important in view of the unavoidable approximations involved in the models of the metal–solvent interaction. Most of these models may appear to be crude approximations of reality. However, it appears that more sophisticated models would not change qualitatively the conclusions on the variation of the metal–solvent distance with charge, provided one considers only situations where the metal–solvent interaction is indeed weak.

Effect on the Calculated Capacitance

The relevance of the variation of d with charge in the calculation of the capacitance will be discussed in this section. For this purpose, we recall Eq. (11), in which the capacitance is expressed in terms of $X_0(\sigma)$ and $X_s(\sigma)$:

$$1/C = 4\pi[X_s(\sigma) - X_0(\sigma)] - 4\pi \partial/\partial\sigma \int_{-\infty}^{+\infty} P(z)\, dz \qquad (11)$$

This expression shows that in addition to the contribution of the polarization $P(z)$, the solution contributes to the capacitance via the variation with charge of the center of mass of $\rho_s(z)$, the excess charge in the solution. In order to understand how the distance of closest approach d for the solvent indeed plays a role in the above expression, let us write $\rho_s(z)$ as

$$\rho_s(z) = -\sigma\delta[z - (d + R_0)] + \tilde{\rho}_s(z) \qquad (33)$$

In this expression, a planar charge distribution peaked on the plane $z = (d + R_0)$, where R0 is the molecular radius, has been isolated. The term $\tilde{\rho}_s(z)$ is the remaining part of $\rho_s(z)$, so that the net charge corresponding to $\tilde{\rho}_s(z)$ vanishes. Then, by writing Eqs. (6) and (10) for the solution side and using the formal splitting in Eq. (33), we get

$$X_s(\sigma) = d + R_0 + \sigma \partial d/\partial \sigma + \tilde{X}_s(\sigma) \tag{34}$$

where $\tilde{X}_s(\sigma)$ corresponds to $\tilde{\rho}_s(z)$. We can now rewrite Eq. (11) as follows:

$$1/C = 4\pi[d - R_0 + \sigma \partial d/\partial \sigma + \tilde{X}_s(\sigma)]$$

$$+ 4\pi\left[2R_0 + \tilde{X}_s(\sigma) - \partial/\partial \sigma \int_{-\infty}^{+\infty} P(z)dz\right] \tag{35}$$

At this level, this expression results only from the introduction of a particular splitting of $\rho_s(z)$ in the exact original expression, Eq. (11). Its structure can be understood by comparing it to the corresponding expression in classical models. In these models, the inner-layer capacitance C_i corresponds to a monolayer of solvent between two charged planes. The first one is the metal surface, so $d = R_0$. An ideal behavior of the metal being further assumed, $X_0(\sigma) = 0$. The first term in brackets then vanishes in the traditional approach. We then define the metal contribution precisely as this term:

$$1/C_m = 4\pi[d - R_0 + \sigma \partial d/\partial s - X_0(\sigma)] \tag{36}$$

and the remaining part of $1/C$ as $1/C_s$:

$$1/C_s = 4\pi\left[2R_0 + \tilde{X}_s(\sigma) - \partial/\partial \sigma \int_{-\infty}^{+\infty} P(z)dz\right] \tag{37}$$

At the present stage these expressions are simple definitions, since only the total capacitance is a measurable quantity. C_m defined in this way has no analogue in the classical analysis of the capacitance. Its definition means also that C_m is not a purely metallic contribution since it contains the effect of the metal-solution coupling through d and its variation with charge.

Metal-adsorbate Interactions

Now we come to the interpretation of C_s. In the classical models and in the absence of specific ionic adsorption, the excess charge in the solution at high enough concentration is assumed to be distributed on the outer Helmholtz plane, located at $(d + R_0)$. Then it follows from the definition of $\tilde{\rho}_s(z)$ in Eq. (33) that $\tilde{X}_s(\sigma) = 0$. If one further assumes that $P(z)$ is limited to a monolayer between $d - R_0$ and $d + R_0$ and splits it as usual into orientational and distortional contributions:

$$P(z) = P^{pol}(z) + P^{or}(z) \qquad (38)$$

and makes the usual assumption $P^{pol}(z) = (\varepsilon - 1)4\pi\sigma/\varepsilon$, where ε is the dielectric constant at fixed orientation of the dipoles, a simple calculation leads to the following result:

$$1/C_s = 4\pi[2R_0 - \partial/\partial\sigma \int P(z)\,dz] = 4\pi\varepsilon/\varepsilon - 1/C_{dip} \qquad (39)$$

where $e = 2R_0$ is the thickness of the monolayer, and C_{dip} is associated with the permanent dipoles. Then C_s defined in this way is exactly equivalent to the inner-layer capacitance considered in the traditional approach, as discussed in Section II.3.

An important consequence of Eqs. (35)–(37) and (39) concerns the analysis of the capacitance at high field. From the experimental curves (Fig. 1), we have seen that $C_i(\sigma)$ exhibits a plateau at sufficiently negative charge. The corresponding capacitance is traditionally denoted K_{ion}. It is usually accepted that K_{ion} corresponds to saturated orientational polarization. Its value is given by the term $4\pi\varepsilon/\varepsilon$ in Eq. (39). It follows from Eqs. (36) and (37) that this limit is actually

$$\lim_{s \ll 0}(1/C_i) = 4\pi[d(\sigma \ll 0) - R_0 - X_0(\sigma \ll 0)] + 4\pi\varepsilon/\varepsilon \qquad (40)$$

[in contrast to the metal–vacuum case, the value $X_0(\sigma \ll 0)$ indeed saturates due to the effect of the solvent through the decrease of d]. From Eq. (40), it appears that there exists in the high-field limit an extra contribution to K_{ion}, which depends on the nature of the metal, at least through X_0. One may say that this extra term is somewhat incorporated in the inner-layer thickness e. We may consider that Trasatti and Valette have made a step in this direction in their interpretation of K_{ion} for solids although their discussion is only based on steric effects. We should say now that both d and X_0 may depend on the orientation of the surface. The presence of the first term on the right-hand side of Eq. (40) will have an important consequence on the estimation of the ratio ε/ϵ. Instead of equating $4\pi\varepsilon/\epsilon$ to the reciprocal value of the capacitance at the plateau, this experimental value should be used now for the whole expression in Eq. (40). Since there is no reason why the extra contribution should be negligible, this will have serious consequences on the value of ε/ϵ and hence on the estimation of the dielectric constant ϵ.

Coming back to the splitting of $1/C$ given in Eqs. (35)–(37), the first task for the theory is to calculate C_m by a model for the metal and for the coupling. Without going into the details of these models, we present in Fig. 12 three determinations of this quantity. They include our work on Ag and the "3D jellium" calculation of Halley and Price with rs corresponding to the value for Cd. We have also plotted a curve from the data of Feldman et al., with r_s corresponding to the value for Hg. Since the original curve of Halley and Price contains the contribution of the solution given by the MSA, we have made the appropriate correction so that the plotted curve corresponds to our definition of $1/C_m$. It differs from the original one only by a constant term. These curves are in general inverted parabolas, with a maximum at a negative charge. Their overall shape results mainly from the variation of $d(\sigma)$, slightly reduced by the monotonic variation of $X_0(\sigma)$. One

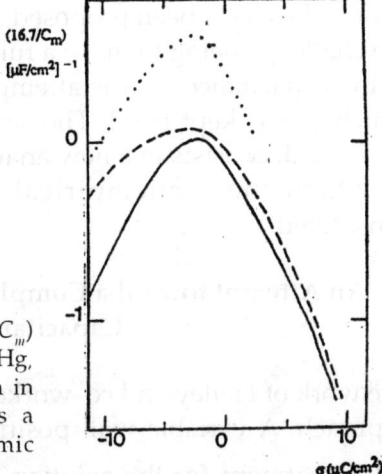

Figure 12. Examples of $(1/C_m)$ curves. ——, Ag; - - -, Cd; ..., Hg, estimated from the data given in Fig. 3. The number 16.7 is a conversion factor from atomic units to $(\mu F\, cm^{-2})^{-1}$.

may also observe that the curve corresponding to the results of Feldman *et al.* is very similar to our curve for Ag. Thus, an agreement on the shape of C_m exists in the literature. If d is maintained fixed, the metal contribution is, of course, totally different, since it is restricted to the variation of X_0 with σ. In that case, it exhibits a monotonic variation with charge.

Some important points may be retained from this discussion: (1) there exists in the capacitance a contribution C_m given in Eq. (36) with no analogue in the classical models; (2) this contribution is largely determined by the variation with charge of the metal–solvent distance; (3) the overall behavior of this distance can be inferred from simple physical considerations in the case of weak metal–solution coupling; and (4) a new analysis of K_{ion} is needed.

Once C_m is known, the next step is to obtain $P(z)$ and $\tilde{X}_s(\sigma)$. *A priori*, this is an intricate problem since the two sides of the interface are coupled, and all these quantities must be calculated self–consistently. At present, two approaches have been proposed. The first one is ambitious, the challenge being to make a fully microscopic calculation of the capacitance as was attempted in the first models in which d was kept fixed. The second one, which we have advocated, consists in a new analysis of experimental data, in which the semiempirical determination of C_s is considered.

An Attempt toward a Complete Calculation of the Capacitance

The work of Halley and co-workers corresponds to the first approach. A variable wall position is considered, and an MSA treatment for the solution is used. Then, $\tilde{X}_s(\sigma)$ is a constant, independent of charge. Physically, this means that the centroid of the excess charge in the solution is assumed to move rigidly with the hard wall for the solvent. This behavior is the direct extension of the classical view, assuming that the ions in the outer Helmholtz plane are always separated from the metal by a layer of solvent.

After their work using a three-dimensional calculation, which was the fourth reported in a series of papers on the capacitance, Halley and Price" reached the rather pessimistic conclusion that the models are far too simple and that efforts in many directions are still required. While this is certainly true for quantitative purposes, we may argue against such a conclusion in several respects. Firstly, the use of the MSA immediately prevents predicting the behavior of the capacitance away from the pzc. Secondly, the law of inverse addition of capacities connected in series is the reason for the inherent difficulty of a direct calculation of the total capacitance as in Eq. (35). Since these capacities (say, C_m and C_s) are of opposite sign and comparable magnitude, the equivalent capacity (say, C_i) can be extremely sensitive to relatively minor variations of

Metal-adsorbate Interactions

the components in the series. It is somewhat regrettable that this limitation, which applies also to the classical models, is seldom discussed in the literature. For example, in the classical models, reasonable values of C_i result very often from a precarious balance between K_{ion} and C_{dip}. Just as an example, in the model proposed by Parsons, a slight change in the ratio of the dipole moment of the clusters to that of the monomers affects so dramatically the balance between K_{ion} and C_{dip} that the capacitance may diverge or become negative. This discussion is also connected with the so-called "Cooper–Harrison catastrophe. This means, in general, that the direct calculation of C may require a high accuracy on each term in Eqs. (35)–(37). Such an accuracy is dictated by the mathematical form of the equation for C and does not correspond to a physical requirement. This also means that it would be better to deal with inverse capacitances. From our point of view, this difficulty makes direct calculation of C not possible at present.

Basis for a Semiempirical Approach

The alternative route indicated in Section 11.5 avoids the problem that we have just discussed. As noted in Section V.3(iii), $P(z)$ has a rather small effect on the calculation of the electronic profile although the converse is not true. Thus, when focusing on the calculation of $X_0(\sigma)$ and $d(\sigma)$, the effect of $P(z)$ can be neglected. More generally, we have checked that C_m given by Eq. (36) can be calculated independently of the state of polarization of the solution. Then, with a given metal–molecule pseudo–potential $V^{Ps}(\mathbf{r}, \mathbf{R})$ and the expression for the van der Waals potential given in the appendix, we calculate the total metal–solvent potential $V_{ms}(Z)$ from Eq. (31). At the same time the coupling term given by Eq. (32) is introduced in the density functional $E[n(z)]$ [see Eq. 04)], and Eq. (30) is solved with a given approximation of $\rho_0(Z - d)$. This procedure, which requires a self-consistent treatment of d and $n(z)$, gives C_m. From C_m and C_i deduced from experiment after correction for the Gouy–Chapman contribution (or at high enough concentration), we obtain a semiempirical estimate of C_s from the relation

$$1/C_s = 1/C_i - 1/C_m \tag{41}$$

This relation is the basis of the new analysis of the capacitance that we have recently proposed. The behavior of C_s obtained in this way can also be compared to the result of model calculations based on Eq. (37). This procedure may finally be used as a check of hypotheses concerning $\tilde{X}_s(\sigma)$. A route similar in concept to the one we propose has been presented by Molina et al. In this work, they estimated Cm from the knowledge of C_i and a model for C_s.

A SEMIEMPIRICAL APPROACH

We proposed an analysis of the capacitance based on the ideas presented above. The case of silver will be discussed first. For this metal, the present model of the electronic structure allows a reliable calculation of C_m. In addition, well-established experimental results for monocrystalline silver electrodes are available.

The Case of Silver

We estimated C_s by using Eq. (41) and the experimental C_i for Ag/NaF and Ag/KPF6. The C_s curves are given in Fig. 13. We first observe a striking similarity of the curves for the two interfaces, more pronounced than in the case of C_i curves. If $C_i(\sigma)$ and $C_i(\sigma)$ differ in magnitude both are simple bell-shaped curves, at least in the range that we considered, -15 µC cm^{-2} < σ < $+15$ µC cm^{-2}. However, an important difference is that while the maximum in $C_i(\sigma)$ is at a slightly positive charge (NaF) or almost at the pzc (KPF$_6$), the

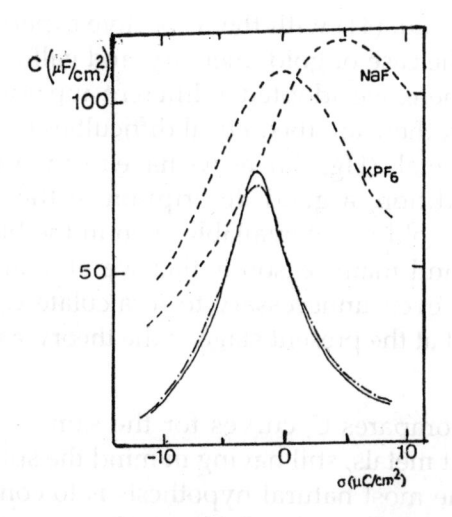

Figure 13. $C_s(\sigma)$ curves for Ag(111)/NaF (—.—) and Ag(110)/KPF$_6$ (———). The experimental C_i (- - -) and given for comparison.

maximum in the corresponding C_s curves is at a negative charge σ–s in both cases.

In the case of silver, for which C_m was calculated, these C_s curves are a direct output of the method. They involve no approximations other than those made in the calculation of C_m. Since they have been obtained independently of any assumption about solvent polarization, they may be considered as the first semiempirical determination of the contribution of the solvent to the capacitance. Since C_s determined in this way cannot be compared directly to experiment, the interesting question is whether or not it helps in rationalizing the role of the solvent. Indeed, we can consider C_s as a quantity from which metal effects have been separated to a large extent. Before discussing $C_s(\sigma)$, it is interesting to consider what may be learned from other interfaces.

Other Interfaces

For other interfaces, we should in principle calculate Cm

and use again Eq. (41) with the respective experimental Ci curves. For the case of gold, mercury, and gallium that we will discuss here, we adopted a different approach for two reasons. First, there are theoretical difficulties encountered with liquid metals (Hg, Ga), as we have mentioned several times. In addition, a good description of the electronic properties of gold is not available even in the bulk phase. The second and main reason is that we ultimately found out that it is even unnecessary to recalculate C_m for each metal, at least at the present stage of the theory, as we detail it now.

If one compares C_i curves for the same solvent but with different metals, still having in mind the splitting into C_m and C_s, the most natural hypothesis is to consider that C_s is the same in all cases, at least as a first approximation. This assumed "universal" behavior is the natural consequence of the way C_s was *defined*, since the effect of the coupling via the distance d and its variation with charge is included in C_m. We stress that the model we used for the coupling does not contain all possible sorts of coupling. For example, it does not include, as it should in principle, an interaction which favors at the pzc a preferential orientation of the water dipole with the negative end toward the metal. We did not include in the calculation of C_m such a "residual interaction," since it adds truly negligible terms in the electronic energy functional. For the same reason as discussed for the effect of solvent polarization in Sections V.3(iii) and VII.4A, such an additional interaction involves energies that are negligible compared to the typical energy experienced by the electrons. On the other hand, since this energy is comparable to that experienced by a dipole, this effect will certainly be required in a model for C_s. It is then known from the early models that the main effect of such a residual interaction is to shift the maximum of the $C_s(\sigma)$ curves, depending on the assumed preferential orientation at the pzc. Then, except for the shift due to this residual interaction, C_s should reflect mainly solvent properties. In this case, since the difference between interfaces is due to C_m, one may estimate Cm for the interface between a metal M' and a solution S' if it is known for another interface between a metal M and a solution S by using −2.1

$$1/C_m(M'/S')=1/C_m(M/S)+[1/C_i(M'/S')-1/C_i(M/S)] \quad (42)$$

which is obtained from Eq. (41) by taking C_s as the same for the two interfaces. Knowing C_m from the calculation for Ag, we obtained the respective curves for Hg and Ga Figure 14 also includes results for Au. We see that while the associated C_i curves in Fig. 1 were dissimilar, all the C_i curves have the shape of an inverted parabola, asymmetric with respect to zero charge, which resulted from the calculations for silver.

A tentative explanation of this behavior of C_m from metal to metal may be advanced from the following physical arguments. We first recall Eq. (36): $1/C_m = 4\pi[d - R_0 + \sigma \, \partial d/\partial \sigma - X_0(\sigma)]$. Since the curves in Fig. 14 have a similar shape, it is sufficient to restrict the discussion to the pzc. Then different values of $1/C_m$ are associated with differences in $d - X_0$. From Fig. 14 we find that the maximum variation of C_m at the pzc corresponds roughly to a difference of about 0.2 Å in the quantity $(d - X_0)$. In our model the values of d and X_0 are coupled. However, for not too different values of d, X_0 depends mainly on the density of conduction electrons. Differences in d result now from differences in the metal–molecule interaction potential $V_{ms}(Z)$, which in turn may be different if we are considering a solid versus a liquid. In our model, $V_{ms}(Z)$ results [Eq. (31)] from the repulsive term V_R and the attractive contribution U_{VdW}. As discussed in Section III.2(iv), V_R for solids stems mainly from the free electrons while it should also include the effect of the ionic cores in the case of liquids. Insofar as U_{VdW} is concerned, its value in the asymptotic region

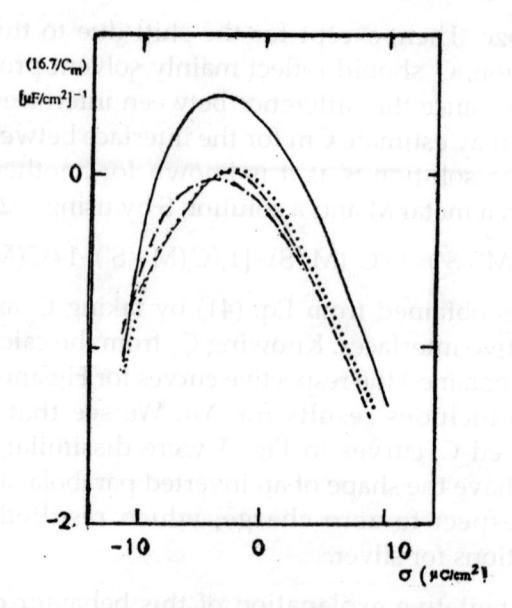

Figure 14. $1/C_m$ curves for different metalas: ———, Hg; —·—, Ga; . . ., Au(210);—, Ag(111). The number 16.7 is a conversion factor from atomic units to $(\mu F\, cm^{-2})^{-1}$.

depends weakly on the nature of the metal, but its value increases with the electron density of the metal at short distances. Since Au and Ag are two solids with very similar electron densities, they are expected to have very close values of X_0 and d. Then similar $1/C_m$ for Ag and Au are expected, just as observed in Fig. 14. If we compare now two liquids, Hg ($r_s = 2.66$) and Ga ($r_s = 2.18$), the discussion is a little more complicated. From the work of Lang and Kohn, we expect a larger value of X_0 by about 0.1 Å for Ga since its electron density is much higher than that of Hg. Then if the values of d were similar, $(d - X_0)$ should be smaller for Ga than for Hg. This is the trend observed in Fig. 14. In order now to reproduce the estimated variation from Hg to Ga, d should be smaller for Ga than for Hg by the remaining 0.1 A. Do we expect such a slight decrease of d from Hg to Ga? An expected larger van der Waals attraction due to a much higher electron density for Ga and a weaker repulsion from the ions due to the smaller ionic radius (0.6 Å for Ga versus 1.1 Å for Hg) are in favor of a decrease in d. However, this is opposed by the expected increase of the repulsion due to the free electrons. A final decrease of about 0.1 A is then not surprising. If we now compare Au or Ag to Hg, since all these metals have similar electron densities, the values of X_0 should also be close. Then the higher value of $1/C_m$ and hence of d for Hg may be attributed to the additional repulsion of the molecules by the metallic ions in the case of a liquid, as mentioned above. These arguments may also explain the position of Ga with respect to Au and Ag, although a detailed classification is not possible.

We stress that this explanation is only tentative and that a detailed analysis of the $1/C_m$ curves should include all terms in Eq. (36). For example, an influence of the term o− adlao− is expected at nonzero charge. In addition, the discussion of the influence of the ionic radii should also take into account some effect due to screening. In this discussion we mainly intend to show that since our definition of C_m incorporates the metal–solvent coupling and the relaxation of the surface dipole, the natural hypothesis attributing the difference in experimental capacities to the effect of the metal can indeed be largely substantiated by the variation of the associated physical properties.

In our opinion, a more important point is the fact that the estimated variation of the physical quantities which determine C_m, say, a variation of $(d - X_0)$ by about 0.2 Å, indicates that very little change in these quantities is needed in order to explain the observed variation of C_m from metal to metal. In addition to the implications of this observation for the interpretation of the specificity of the capacitance curves for various metals, it also means that a quantitative account of this behavior can be very demanding. Bearing in mind that $\Delta(d - X_0) \cong 0.2$ Å, if one recalls that the water molecule is often modeled by a sphere with a radius of about 1.5 Å, it is clear how accurate the calculations must be in order to account for such a slim difference between metals.

Effect of Temperature

A very interesting consequence of the previous analysis of the splitting of the capacitance into C_m and C_s is found in examining the temperature dependence of the solvent contribution. From the classical data of Grahame for Hg/NaF (see Fig. 2), the effect of temperature on Cs was estimated by using Eq. (41) and C_m^{-1} for Hg given in Fig. 14, assumed to be temperature independent.† Therefore, C_s curves (Fig. 15) at different temperatures are just another representation of the original C_i curves. However, the difference between the two sets of curves is striking. In contrast to the complicated appearance of the original curves, the behavior of C_s is rather trivial. At a given temperature, $C_s(\sigma)$ is a simple bell–shaped curve. A maximum in the C_s curve due to the maximum randomness of dipole orientations is naturally observed at all temperatures. The interplay of the variations of C_m and C_s results in the occurrence of a hump in C_i at a slightly positive charge and only at low temperature. At a given charge now, increasing T reduces the magnitude of C_s. The so–called temperature–invariant point is also much less marked in the $C_s(T, \sigma)$ curves in Fig. 15 than in the corresponding C_i curves. The rapid drop of C_s observed

---

† This is the usual assumption. Arguments for this can be given. Here, we only mention that since in our approach the behavior of C_m is dominated by that of d, which depends little on temperature, the assumption that C_m also does not depend on temperature is quite reasonable.

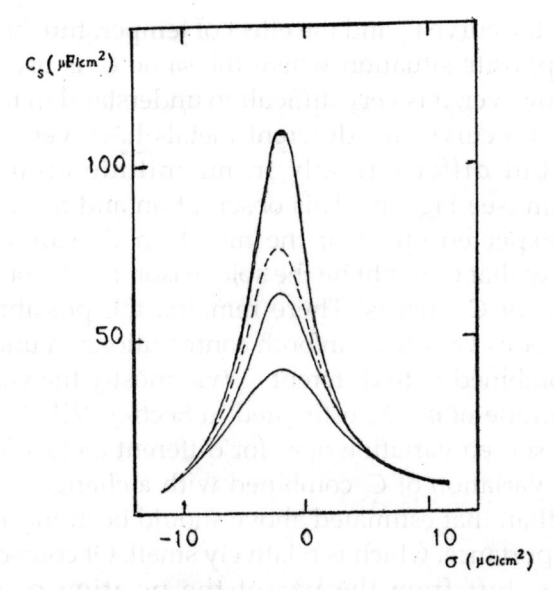

Figure 15. Cs curves for Hg/NAF at different temperatures. From top to bottom: T=0, 25, 45, and 85°C. The dashed curve corresponds to C_s(Ag/NAF).

away from the maximum may simply reflect saturation of the polarization.

For the aqueous solutions considered above, another important point is the systematic occurrence of the maximum of C_s at a negative charge. As substantiated by entropy curves, this observation is in accordance with elementary considerations concerning the orientation of polar molecules, with preferential orientation of the negative end toward the metal. This important feature follows naturally once the contribution C_m is separated from the inner–layer C_i curve and provides a very simple reconciliation of entropy and capacitance data.

We stress here that the working hypothesis which underlies the discussion above is that C_s depends mainly on the solvent. We observed that the conclusions about the shape of the curve C_s and the effect of temperature hold also in the opposite situation where the same C_s is used in all cases. However, it is very difficult to understand in this case why the Cs curves for different metals have very similar shapes but differ strongly in magnitude around the maximum (see Fig. 3b). This observation and the remarks on the expected effect of the metal on C_m rule out the possibility that C_s might be the sole reason for the observed variation of C_s curves. There remains the possibility of simultaneous variations in both contributions. A unique C_s curve combined with different C_s (via mostly the variation in magnitude of $d - X_0$ estimated in Section VIII.2) results in the observed variation of C_s for different metals. Thus, a possible variation of C_s combined with a change in $d - X_0$ smaller than that estimated above should be at most of the same importance, which is relatively small. Of course, there is also a shift from the pzc of the position σ_m of the maximum of C_s which may depend on the nature of the metal. However, a metal. dependent value of σ_m should be in a very narrow range. Indeed a large change in the maximum of $1/C_m$ is required in order to have a significant change of the position of the maximum of the C_s curves deduced from the C_i curves. Such a change would hardly be compatible with the mechanism which determines $1/C_m$, via $d - X_0$, as a function of charge.

Of course, one cannot strictly exclude the case where both C_s and C_m change considerably with the nature of the metal in some unpredictable way. However, given the physical meaning of C_s and C_m as we defined them, this hypothesis is very unlikely. It seems very reasonable that, in reality, both C_s and C_m change with the nature of the metal, but only slightly. What should also be made clear is that while dissimilar C_i curves may result from minor changes in C_s or C_m or both, the splitting of a given C_i curve into C_s and C_m results always in stable curves. This ultimately means that the sensitivity of our analysis to small

Metal-adsorbate Interactions

variations in C_s and C_m, as a consequence of the law of inverse addition of capacitances connected in series, is an advantage and not a drawback as it may appear at first sight.

Role of the Solvent

As a final illustration of the previous analysis, we consider now the effect of changing the nature of the solvent. Details can be found in our recent paper. In this paper, we developed arguments the same showing that, for qualitative purposes, one may first use C_m in analyzing C_i data for nonaqueous solvents at the mercury electrode. From inner-layer capacitance curves $C_i(\sigma)_{Hg/sol}$ for different solvents, we defined the solvent contribution as

$$C_s^{-1}(\sigma)_{sol} = C_i^{-1}(\sigma)_{Hg/sol} - C_m^{-1}(\sigma)_{Hg} \tag{43}$$

where $C_m^{-1}(\sigma)$ is the curve for Hg given in Fig. 14. $C_s(\sigma)$ obtained by using Eq. (43) is obviously a very indirect estimate of the solvent contribution so that details of the $C_s(\sigma)$ curves may not be reliable. However, the curves obtained in this way and shown in Fig. 16 present some features which are very unlikely to be artifacts of the method.

While their overall shape is the same, the $C_s(\sigma)$ curves depend, as expected, on the nature of the solvent in terms of their relative magnitudes and the charge σ_m corresponding to their maximum. For formamide, water, propylene carbonate, ethylene carbonate, and dimethyl sulfoxide, σ_m is negative. It is almost zero for dimethylformamide and acetone and positive for methanol and acetonitrile.

We can now discuss the kind of information one may derive from σ_m. As is well known, models based on a simple

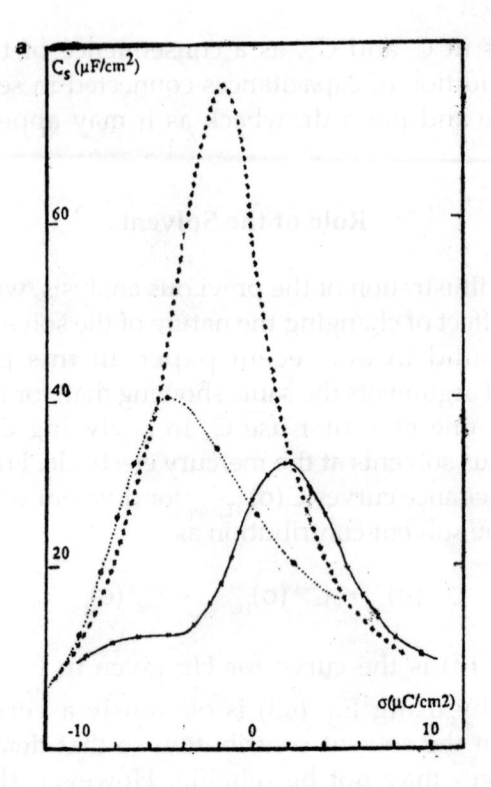

Figure 16. $C_s(\sigma)$ curves: (a) water (x x x), methanol (-x-), and formamide (●...●); (b) dimethylformamide (—.—), dimethyl sulfoxide (—), acetonitrile (+ + +), propylene carbonate (—), ethylene carbonate (-●-), and acetone (...).

mechanism for the solvent orientation as a function of charge predict that the maximum of $C_i(\sigma)$ corresponds to the maximum randomness of the orientation of the dipoles. In this class of models, the sign of the charge at which the theoretical $C_i(\sigma)$ is maximum is directly related to the assumed preferred orientation of the dipoles at the pzc. The maximum of the capacitance is also at the same charge as the maximum of the entropy. The latter is thought of as a more direct probe of solvent orientation, although the analysis of entropy data also requires a model. However,

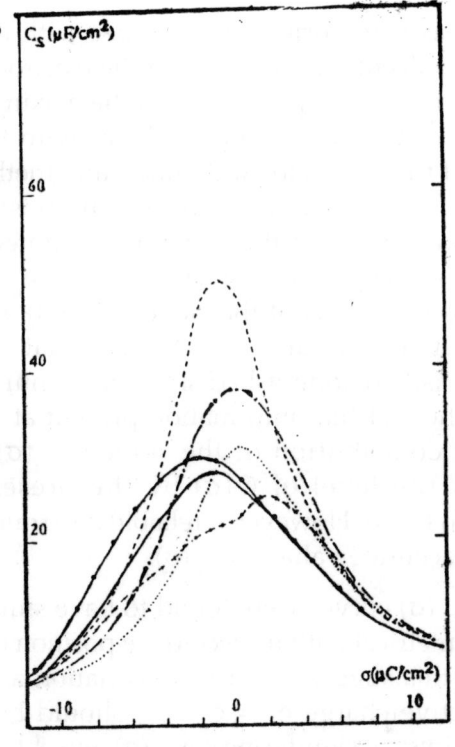

Figure 16. (*Continued*)

if one considers a more complex mechanism such as, for example, the presence of water clusters, the capacitance hump in the theoretical C_i does not correlate with the maximum of the entropy.

In the class of models attributing the hump to the orientation of the solvent dipoles, this absence of correlation between capacitance and entropy data led to the conclusion that maxima or humps in the capacitance curves give no information on solvent orientation at the pzc. Since no single criterion allows a definitive conclusion, the preferred orientation has to be assessed by comparing results from different experiments. On the basis of such a coherent set of observations, most of the solvents in Fig. 16 are usually considered as being oriented at the pzc with the negative end of the molecular dipole toward the electrode. It can be seen in Fig. 16 that the position σ_m of the maximum of $C_s(\sigma)$ is negative or very close to zero. Thus, with the possible exception of the curves for acetonitrile and methanol, these curves show a correlation between the position of the maximum of $C_s(\sigma)$ and the generally accepted preferred orientation at the pzc. The precise value of σ_m depends in this method on details of the slopes of $C_i(\sigma)$ and $C_m(\sigma)$ at moderate charges. In any case, the fact that $C_s(\sigma)$ curves exhibit an easily distinguished single maximum illustrates very well the fact that information present at the level of the, "real" contribution of the solvent $C_s(\sigma)$ is simply masked at the level of $C_i(\sigma)$ by the presence of the contribution $C_m(\sigma)$. However, a reliable determination of o–m requires accurate values of $C_m(\sigma)$.

Since $C_s(\sigma)$ curves were found to have similar shapes, almost symmetrical with respect to the position of the single maximum, this suggests that their variation as a function of the effective charge $\sigma^* = \sigma - \sigma_m$, should be examined Given the physical significance of $C_s(\sigma)$, we obtained the ————————————————— of the charge σ^* from Eq. (39) as

$$C_{\text{dip}}^{-1}(\sigma^*) = 4\pi e/\varepsilon - C_s^{-1}(\sigma^*) \qquad (44)$$

where C_{dip} is associated with the orientational polarization of the solvent, while the distortional polarization is related to $4\pi e/\varepsilon$ as usual. For the reason discussed below, we took the value $\varepsilon = 2$ for all solvents, and e was taken as equal to the molecular diameter determined in such a way as to have a fixed packing fraction η in the bulk for all solvents: $\eta = \pi\rho e^3/6 = 0.45$, where ρ is the bulk density. $C_{\text{dip}}(\sigma^*)$ curves obtained in this way are plotted in Fig. 17, from which the following observations can be made.

Around $\sigma^* = 0$, the variation of $C_{\text{dip}}(\sigma^*)$ for different solvents is considerably reduced, when compared to that of $C_s(\sigma)$ in Fig. 16. This confirms the dominant role of molecular size advocated many times (see, for example, the review by Payne). At moderate σ^* the curves for dimethyl sulfoxide, dimethylformamide, propylene carbonate, ethylene carbonate, and acetone merge almost into a single curve, while those for H_2O, methanol, and formamide are clearly separated. Note that the curve for acetonitrile has an intermediate position. At higher magnitudes of $0'^*$, polarization depends on the solvent.

The behavior of $C_{dip}(\sigma^*)$ for nonzero σ^* reflects the role of intermolecular interactions. A detailed discussion has been presented in our recent paper.[155] The sequence near $\sigma^* = 0$ can be understood qualitatively, from the competition between the orienting effect imposed by the external field and the effect of intermolecular interactions which oppose it. With regard to the field, simple considerations show that $C_{dip}(\sigma^*)$ is proportional to $(e/\mu)^2/\partial(\cos\theta)/\partial\xi$, where f is a reduced quantity defined by $\xi = \mu E^*/kT$. $E^* = E - E^r$ is the effective field, which reflects the competition between the applied field E and the residual field E^r, μ is the dipole moment, and θ is the angle that the dipole makes with the normal to the electrode surface.

For aprotic solvents, intermolecular interactions are mainly dipolar. An approximate measure of the strength of dipolar interactions is the ratio $v_{dd} = (\mu^2/kTe^3)$. For these aprotic solvents with similar values of the dipole moment and of the radius, v_{dd} and the coefficient $(e/\mu)^2$ do not change much (see Table 4), so that similar values of C_{dip} are expected. Indeed, the values of C_{dip} differ little for these solvents at moderate charges with the exception of acetonitrile. This would indicate that deviations from the simple picture given above (polarizability, nonspherical shape, etc.) are relatively unimportant near $\sigma^* = 0$. The position of acetonitrile, which has a relatively smaller radius, can be understood if one considers the effect of the radius on C_m. Inspection of Table 4 shows that the position of acetone results from the combined effect of its larger value of $(e/\mu)^2$ and its smaller value of v_{dd}.

Table 4
Solvent Physical Parameters

Parameter	H$_2$O	MeOH	FA	AC	EC
e(Å)	3	3.86	3.83	4.71	4.55
μ(D)	1.84	1.66	3.66	2.88	4.87
$(e/\mu)^2$ $(\mu/D)^2$	2.65	5.40	1.09	2.67	0.87
v_{dd}	3.02	1.15	5.75	1.91	6.07
$\varepsilon/4\pi e$(μFcm^{-2})	5.89	4.58	4.61	3.75	3.88

Parameter	PC	DMSO	DMF	ACN
e(Å)	4.94	4.66	4.78	4.21
μ(D)	4.94	3.96	3.82	3.92
$(e/\mu)^2$ $(\mu/D)^2$	1	1.38	1.56	1.15
v_{dd}	4.88	3.73	3.22	4.96
$\varepsilon/4\pi e$(μFcm^{-2})	3.57	3.79	3.70	4.20

For associated solvents, H bonds, corresponding to energies in the range 5–20 kT, that is, larger than $v_{dd}kT$, dominate intermolecular interactions. Thus, the molecules of these solvents are more difficult to orient than those of unassociated solvents and hence should have smaller values of $\partial(\cos\theta)\partial\xi$. Except for formamide, they also have larger values of the ratio $(e/\mu)2$. As observed in Fig. 17, $C_{dip}(\sigma^*)$ should be larger for water, formamide, and methanol than for aprotic solvents. At higher charges, the competition between the applied field and intermolecular

Metal-adsorbate Interactions

interactions makes the rise of $C_{dip}(\sigma^*)$ specific to each solvent. A detailed discussion is, however, difficult since the precise value of $C_{dip}(\sigma^*)$ at these charges is very sensitive to uncertainties in e/ε and $C_s(\sigma^*)$.

In conclusion, we may say that when $C_m(\sigma)$ is extracted from $C_i(\sigma)$, a clear picture of the behavior of different solvents emerges. It can be rationalized largely in terms of their physical parameters (dipole moment, molecular size) and the nature of intermolecular interactions. This is to be compared with the often adopted classification of solvents into different groups according to the shape of $C_i(\sigma)$. In addition, the preferred orientation of the solvent dipoles at the pzc can be deduced from the position of the maximum of C_s.

Simple Model for Cs

The features discussed above and the overall behavior of C_s with charge and temperature suggested that a simple model for dipole orientation in an electric field be tried. We indeed found that a model of the Langevin type for the polarization gives an excellent account (see Fig. 18) of the semiempirical C_s for water. In this model, the polarization is given by

$$P = N(\mu \cos\theta) = N\mu[\coth(x) - 1/x]; \quad x = \mu E_m/1kT \quad (45)$$

where N is the density of dipoles, and μ is their dipole moment (6.1 10^{-30} C m, or 1.84 D, for water). The mean field orienting the dipoles was taken as

$$E_m = 4\pi\sigma^*/\sigma \quad (46)$$

where ε is an adjustable parameter, and $\sigma^* = \sigma - \sigma_m$ takes into account any preferential orientation of the dipoles at the pzc. The resulting capacitance is

$$1/C_s = 4\pi e - (4\pi N_s\mu^2/kT)[1-\coth^2(x) + 1/X^2] \quad (47)$$

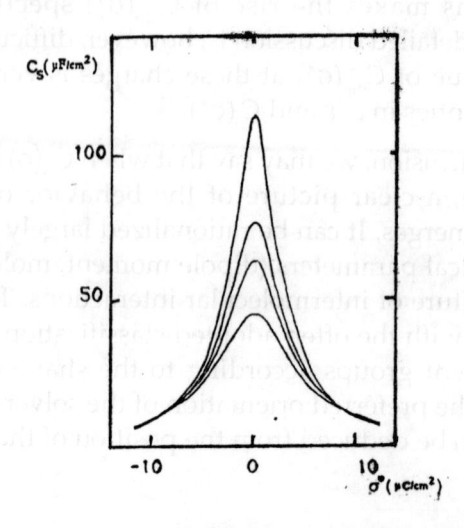

Figure 17. Theoretical $C_s(\sigma^*)$ curves with a one-parameter model; σ^* is the effective charge on the electrode. From top to bottom: T = 0, 25, 45, and 85°C.

where e is the thickness of the solvent layer, and N_s is the number of dipoles per unit area. Considering a monolayer of thickness $e = 2R_0 = 3$Å and hexagonal packing, one has $N_s \cong 1.2 \times 10^{19}$ m^{-2}. The values of C_s at the maximum were then fitted at different temperatures with the help of e, giving:

T(°C)	0	25	45	85
ε	14.35	13.31	12.60	11.40

These values of ε are in the range of the dielectric constant at weak field and decrease with temperature as expected. Note that ε may incorporate various effects neglected in the model, such as lateral interaction, image forces, or H bonds. See also the discussion of Marshall and Conway for a more refined treatment of this kind of model.

As detailed in our previous work, a better fit of the saturated value of C_s in Fig. 17 is obtained by introducing into the model the polarizability of the molecule. In contrast to Eq. (47), the saturated value of $1/C_s$ is now $4\pi e/\varepsilon_\infty$. We emphasize that the parameters in this "improved" model were given in Ref. 60 for the purpose of illustration and are not to be taken literally, especially in the expression of the dielectric constant $\varepsilon(\sigma)$. The overall treatment of the model is also not self-consistent. However, the important point is that the saturated value $C_s(|\sigma| \gg 0)$ given in the model by the term $4\pi e/\varepsilon_\infty$ (usually denoted by K_{ion}) is obtained with e_∞ close to $e_\infty = 2$. Incidentally, this value of ε_∞ happens to correspond to the value obtained from optical measurements. Such a small value of ε_∞ is obtained because the contribution of the metal was extracted from the experimental C_i, so that $C_s(|\sigma| \gg 0)$ is smaller than the classical value of about 17 µF cm^{-2} (recall the discussion in Section VII.2). While the approximations in the theory make it difficult to state that this value should be taken as definitive, this point strongly suggests that the accepted value of the "dielectric constant of the inner layer" should be examined.

Finally, our analysis of C_s remains with the traditional view: we deal with capacitances corrected for the Gouy–Chapman contribution, consider a monolayer model for C_s and a local dielectric constant, and so on. This may appear rather strange if we recall the important developments in statistical mechanics which provide, in principle, the tools for a more rigorous treatment of the model. These developments also cast some doubt on the validity of conclusions drawn from molecular models which do not incorporate an accurate description of the short-range interfacial structure (see the recent work of Torrie et al and references therein). We are aware of the fact that the simple calculation given above does not constitute rigorous treatment of the model. Moreover, the introduction of unknown parameters is related to physical effects, such as dipole–dipole or ion–dipole interactions, that can be dealt with properly only by an elaborate statistical mechanical treatment. The practical reason for our choice of a less rigorous and hence much simpler treatment was the observation that the shape of C_s is simple. One may guess that a rigorous treatment of the model will possibly remove the need for adjustable parameters or clarify their physical significance and precise values. Thus, it will be interesting to investigate in the future the possibility that an elaborate treatment of the solution ultimately gives the same results as a less sophisticated one (say, a Langevin function for the polarization) even if only on a formal level.

4

Electrocatalytic Oxidation of Oxygenated Aliphatic Organic Compounds

INTRODUCTION

The investigation of electrocatalytic processes involved in the electrooxidation of organic compounds became a subject of growing interest at the beginning of the 1960s. A tremendous effort was undertaken to develop fuel cells, which theoretically have the ability to directly convert the chemical energy of hydrocarbons into electrical energy without the limitations due to Carnot's theorem (i.e., with a high energy efficiency), an advantage which is particularly attractive for autonomous power sources. In such applications the rate of conversion of the organic compound by oxidation into carbon dioxide has to be as high as possible, in order to obtain the maximum energy available from the fuel and thus the maximum efficiency. However, the incomplete oxidation of organic compounds is of great interest as well, if one considers the potential applications for organic electrosynthesis, so that many attempts have been made in this field as well. Rigorous control of the experimental conditions may allow such reactions to become highly selective, leading to the development of industrial processes based on electrochemical reactions for the production of chemicals.

These two aspects, namely, energy conversion and chemical production through the electrocatalytic oxidation of oxygenated organic compounds, will be discussed in this chapter. Emphasis will be given to detailed kinetic studies of electrode reactions. Conversely, the large field of organic polarography and of electroanalytical methods in organic chemistry, which is mainly oriented toward electrochemical reactions whose rates are mass-transfer controlled, will not be examined here. Oxygenated aliphatic molecules with one to six carbon atom" in length, including monosaccharides, with single or multifunctional groups (alcohols, aldehydes, ketones, carboxylic acids) will be considered here. Most of these molecules are highly soluble in water, so hat their electrochemical transformation can be studied in aqueous electrolytes. Organic and nonaqueous solvents will not be discussed here, nor will hydrocarbons or aromatic compounds, which are beyond the scope of this chapter. Some data on the oxidation of hydrocarbons may be found in other review papers.

Electrocatalysis, which plays a key role in the mechanism of electrochemical reactions at solid electrodes, can be defined its the heterogeneous catalysis of' electrochemical reactions by the electrode material. Thus, the role of the nature and structure of the electrode is emphasized. For example, depending on the nature of the electrocatalytic metal or alloy, the rate of oxidation of' methanol to carbon dioxide, expressed as the exchange current, density, varies over several orders of magnitude, from 10^{-10} A cm^{-2} for iridium to 10^{-4} A cm^{-2} for platinum-iron alloys.

The structure of a given electrocatalytic metal, such as platinum, also greatly influences the reaction rate, as clearly seen using electrodes of well-defined and controlled structure (such as low-index single crystals). Since the pioneering work of Clavilier *et al.*, who were able to control the preparation of platinum single crystals and to characterize the electrode surface by electrochemical techniques, several papers have been published that definitively show that electrocatalytic reactions such as the oxidation of' methanol, formaldehyde, formic acid, and carbon monoxide are structure-sensitive.

Other ways of controlling the catalytic properties of the electrode surface, such as metal alloying and underpotential deposition (upd) of foreign metal adatoms on noble metal electrodes, will be considered. In these two cases, enhanced catalytic effects have been observed, insofar as the electrooxidation of small organic molecules on platinum electrodes is concerned.

Most of the electrode materials considered in this chapter are pure noble metals (Pt, Rh, Pd, Au, Ir, Ru, etc.), either smooth, such as single crystals or preferentially oriented surfaces, or rough, such as polycrystalline metals. They can also be alloyed with other noble or non-noble metals or modified by upd of foreign metal adatoms (Ag, Bi, Cd, Cu, Pb, Re, Ru, Sri, Tl, etc.).

In electrocatalysis, as in heterogeneous chemical catalysis, the substrate affects the reaction kinetics mainly through the adsorption or reactants, reaction intermediates, and products. Therefore, special emphasis will be put on the adsorption of organic molecules at noble metal electrodes. Since the last reviews on the subject, by Damaskin *et al.* and by Breiter (in this series), tremendous progress has been made in the understanding of adsorption processes at solid electrodes, particularly due to both the use of well-controlled surfaces and the development of in situ spectroscopic techniques, such as infrared (IR) reflectance spectroscopy, which can provide information on the structure of adsorbed species at the molecular level.

Weakly bonded species, acting very often as reactive intermediates, and strongly bonded species, which are irreversibly adsorbed at the electrode surface and which block the electrode active sites (poisoning species), have both been identified by IR reflectance spectroscopy. It is now recognized from spectroscopic measurements that the electrocatalytic poisons formed during the electrooxidation processes of many small organic molecules are adsorbed CO (both linearly and bridge-bonded to the surface), and not COH as previously believed. It is now understood that relatively long spectral accumulation times favor the formation of strongly bonded intermediates, which accumulate at the electrode surface and displace practically all the weakly bonded species. Therefore, as previously pointed out, whereas the poisoning species are rather easily detected by spectroscopic methods, it is much more difficult to obtain information on weakly bonded species and on reactive intermediates. However, new improvements in the measurement techniques, particularly in the case of mass spectroscopy and real-time IR reflectance spectroscopy, have begun to provide some solutions to this difficult problem. Mass spectroscopy, for instance, has been used to analyze "on-line" the volatile products generated at a porous platinum catalyst supported on a Teflon membrane. As only desorbed species are observed, this implies that they come from weakly adsorbed species." In the case of methanol adsorption, the structure of the adsorbed species was suggested to be CHO or COH. With recent IR spectroscopic techniques, very high sensitivity can be achieved (10^{-4} to 10^{-6} in relative changes of' absorbance), and shorter adsorption times (on the order of a few seconds) can now be monitored. It has therefore become possible to investigate the early stages of organic adsorption and to observe the time evolution of the IR spectra of the adsorbed species. Similarly, by combining Fourier transform infrared reflection-absorption spectroscopy (FTIRRAS) and electrochemical thermal desorption mass spectroscopy (ECTDMS), it has been possible to detect both reactive intermediates and poisoning intermediates (of the CO type) during ethanol adsorption and oxidation at platinum 21 electrodes.

In this chapter, as mentioned above, detailed reaction mechanisms will be discussed, particularly those concerning the electrocatalytic oxidation of C_1 molecules (CO, HCOOH, HCHO, CH_3OH), for which the extensive work of the last decades has led to the unambiguous identification of the reactive intermediates and poisoning species. Purely electrochemical methods (voltammetry, chronopotentiometry, chronoamperometry, coulometry, etc.) will be discussed first, as they are the basis of any electrochemical stud~ of adsorption and electrooxidation reactions. Then, the newly developed spectroscopic and microscopic techniques will be particularly emphasized, as they have become unique tools, not only for studying in situ the structure of the electrode/electrolyte interface at the molecular level, but also for investigating the nature and the structure of adsorbed intermediates. The quantitative analysis of the reaction products by modern analytical techniques (gas and liquid chromatography, mass spectroscopy, nuclear magnetic resonance, Fourier transform infrared spectroscopy, etc.) will also be discussed. These methods provide information on the mass balances and the faradaic yields, which are essential for correctly writing the reaction mechanisms.

In the last part of the chapter, selected examples, all of them dealing with the electrocatalytic oxidation of small aliphatic organic molecules, will be presented. The electrochemical reactivity of different functional groups (alcohols, aldehydes, ketones, carboxylic acids) at various catalytic anodes and the influence of chain length in a homologous series of molecules (e.g., the aliphatic monoalcohols from C_1 to C_6 and of the position of the functional group in structural isomers (e.g., the 4-butanol isomers) will be successively examined. The electrooxidation of multifunctional molecules, particularly dialcohols, dialdehydes, and diacids, will be considered as well. Finally, the oxidation of polyols (ethylene glycol, glycerol, sorbitol, etc.) and monosaccharides (glucose, fructose, etc.) will be discussed, particularly in terms of reaction mechanisms and selectivity of electrode reactions.

EXPERIMENTAL METHODS

The investigation of the kinetics of electrocatalytic reactions, such as the electrooxidation of organic compounds, requires the use of several complementary methods, both at the macroscopic level and at the molecular level. The reactivity of the organic molecules, the activity of the electrode materials, and the overall reaction mechanisms are conveniently studied by electrochemical methods, among which voltammetry is particularly suitable.

Adsorbed intermediates, which play a key role in electrocatalytic reactions, may be investigated in situ either by electrochemical methods (e.g., pulse voltammetry, coulometry, impedance measurements), by spectroscopic methods (e.g., UV-visible, IR, Raman, mass spectroscopy), or by other physicochemical methods (e.g., radiometric techniques, quartz microbalance, scanning tunneling microscopy).

The nature and structure of the electrode, which determine the electrocatalytic properties of the surface, have to be controlled and may be analyzed by convenient ex situ methods, such as electron microscopies (e.g., SEM, TEM), electron diffraction techniques (e.g., LEED, RHEED), and electron spectroscopies (e.g., XPS, ESCA, AES).

To determine the mass balance and the faradaic yields, the reaction products and by-products must also be quantitatively analyzed by analytical methods, such as chromatographic techniques (e.g., GC, HPLC) and other quantitative analytical methods (e.g., FTIR, NMR, MS).

The different experimental techniques now available to investigate electrocatalytic reactions will be outlined briefly in this section focusing on the type of information obtained by each technique. For more details on the techniques themselves, the reader will be referred to specialized books or reviews.

Investigation of the Overall Reaction Kinetics by Electrochemical Methods

The usual kinetic parameters of an electrochemical reaction, such as the exchange current density i_0 (or the standard rate constant k^0), the reaction orders p_i, the stoichiometric number v, the transfer coefficient a (or symmetry factor), the standard heat of activation ΔH_0^*, and the rate-determining step, are usually determined by steady-state methods (direct potentiostatic or direct galvanostatic methods).

However, for most electrocatalytic reactions involving small organic aliphatic compounds, self-poisoning phenomena arise from the dissociative chemisorption of the molecule. This leads to a blockage of the electrode active sites, so that the current intensity (which is proportional to the electrode area, that is, to the number of active sites), at a given constant applied potential, falls rapidly to small values. Therefore, only by using transient methods on a time scale far shorter than the usual time scale of adsorption processes (i.e., usually a few seconds or a few tens of seconds) and/or by allowing the electrode surface active sites to be regenerated periodically by oxidizing the blocking species at higher potentials, kinetic information on such electrocatalytic reactions can he obtained.

Among transient methods, voltammetry (which is a potentiody-namic method) is one of the most popular electrochemical methods for the investigation of the overall kinetics of electrocatalytic reactions, since it combines the advantages of short time scales (at this sweep rates up to a few thousand volts per second) and large potential variations (able to oxidize the adsorbed poisons). The method consists in applying to the electrode under study a line potential sweep between two potential limits, E_a and E_c, an recording the current flowing through the electrode as a functio of the applied potential, which is equivalent to recording current versus time curves (Fig. 1). A single potential sweep is applied i linear potential sweep voltammetry (LPSV), whereas repetitive triangular waveforms are used in cyclic voltammetry (CV). This method was first used by Will and Knorr to study electrocatalytic processes at noble metal electrodes, namely, the adsorption of hydrogen and oxygen at platinum.

The analysis of adsorption processes in the case of reversible and irreversible charge transfer reactions was discussed by Srinivasan and Gileadi and later by Angerstein- Kozlowska *et al.* and Alquié-Redon *et al.* The case of electron transfer reactions coupled to diffusion processes was also considered and treated theoretically, but is not directly relevant to electrocatalytic reactions, since, in the latter cases, electrode processes are mostly controlled by adsorption rather than by diffusion.

The voltammetric response $I(E)$, that is, the current intensity (I) versus potential (E) curve, which is called a "voltammogram," displays a current peak I_p at a given potential E_p. This peak results from a depletion of the electrode surface in the electroactive species as a consequence of a limited rate of either mass transfer or the

Electrocatalytic Oxidation of Oxygenated... 113

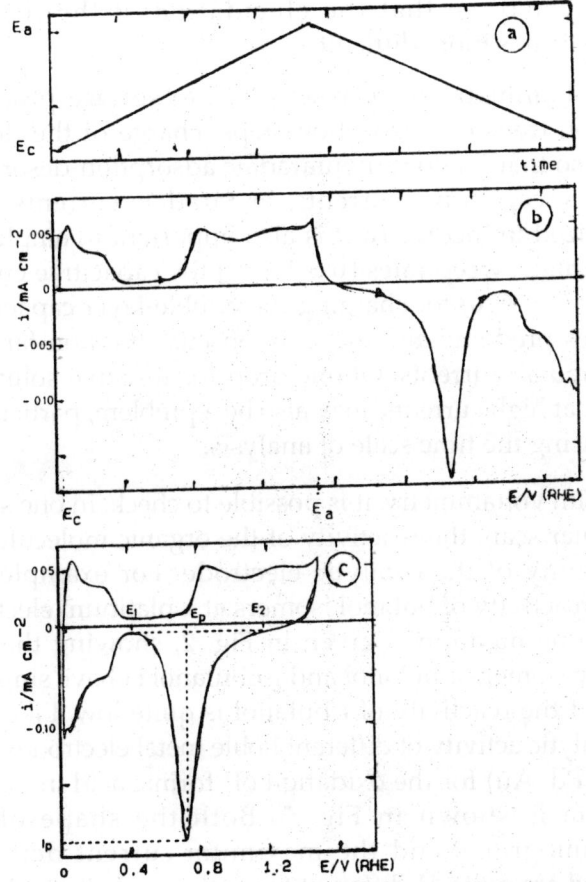

Figure. 1 Schematic principle of cyclic voltammetry, illustrated by voltammogram of a Pt electrode in 0.1 M $HClO_4$ (25ºC, v = 50 mVs^{-1}). (a) Cyclic potential sweep; (b) cyclic current response versus time; (c) cyclic voltammogram.

adsorption process (Fig. 1). The analysis of the peak characteristics (E_p, I_p) as a function of the potential sweep rate $v = dE/dt$ allows, in the simplified case of a single kinetic control (either by diffusion or by adsorption), diagnostic criteria to be established for the reaction mechanisms (Table 1). However, for most cases where mixed kinetic control occurs, the analysis is much more complicated, so that the elucidation of the reaction mechanism is more difficult.

Voltammetry is very sensitive to surface processes (electron transfer, adsorption steps, charge of the double layer), so that it allows nonfaradaic adsorption-desorption currents, capacitive currents, or faradaic currents to be recognized immediately. It is also important to emphasize that at high sweep rates ($v > 1$ Vs^{-1}) the capacitive current density ($i_c = C_d v$) for charging the double-layer capacity C_d becomes predominant and may obscure the non-faradaic and faradaic currents. Ohmic drop for resistive solutions, and/or at high currents, may also be a problem, particularly in reducing the time scale of analysis.

With voltammetry, it is possible to check, in one single triangular scan, the reactivity of the organic molecule and the activity of the catalytic electrode. For example, the electroreactivity of butanol isomers at a platinum electrode in alkaline medium is given in Fig. 2, showing that the primary isomers n-butanol and isobutanol behave similarly and that the reactivity of t-butanol is quite low. Likewise, the catalytic activity of different noble metal electrodes (e.g., Pt, Rh, Pd, Au) for the oxidation of' formic acid in neutral medium is shown in Fig. 3. Both the shape of the voltammograms and the maximum current densities obtained are quite 41 different for each metal electrode.

Voltammetry can be carried out at slow potential sweeps (v 10 mVs^{-1}) to obtain quasi-steady-state current-potential curves $I(E)$, which allow kinetic parameters to be determined. Practical criteria for the quasi-steady-state character of the polarization curves are that the voltammogram does not vary with the sweep rate v and that forward and backward sweeps are nearly superimposed. This means that practically no poisoning phenomena are 42 occurring in the course of the reaction.

Table 1
Peak Characteristics of Voltammograms in the Case of a Simple First-Order Transfer Reaction Controlled Either by Adsorption or by Semi-Infinite Linear Diffusion

	Diffusion control	Adsorption control
	Peak potential E_p	
Reversible transfer	$E^0 + \dfrac{0.0285}{n}$	E^0
Irreversible transfer	$E^0 + \dfrac{RT}{\alpha n_a F}\left[\ln\left(\dfrac{\alpha n_a F}{RT} D_i \dfrac{v}{k_s^2}\right)^{1/2} + 0.78\right]$	$E^0 + \dfrac{RT}{\alpha n_1 F}\ln\left(\dfrac{\alpha n_a F v}{RT k_s}\right)$
	Peak current I_p	
Reversible transfer	$0.466 nF\left(\dfrac{nFD_i}{RT}\right)^{1/2} C_i^0 \sqrt{v}$	$0.25\dfrac{n^2 F^2}{RT}\ln[N_{ads}]v$
Irreversible transfer	$0.496 nF\left(\dfrac{\alpha n_a FD_i}{RT}\right)^{1/2} C_i^0 \sqrt{v}$	$0.368\dfrac{\alpha n^2 F^2}{RT}[N_{ads}]v$

Of the other transient electrochemical techniques, only a few have been considered for the study of the kinetics of electrocatalytic reactions. Among them, pulse techniques, such as chronoamperometry and chronopotentiometry, are the most used. In principle, pulse techniques, and particularly chronoamperometry (in which a potential-step perturbation is applied to the working electrode), are better suited than voltammetry for the study of electrode kinetics, since only time is considered as the experimental variable and since all other potential-dependent quantities (rate constants, electrode coverage, double-layer capacity, etc.) are kept constant at a fixed potential. However, due to the complexity of the reaction mechanisms, kinetic data remain somewhat difficult to obtain with these pulse techniques.

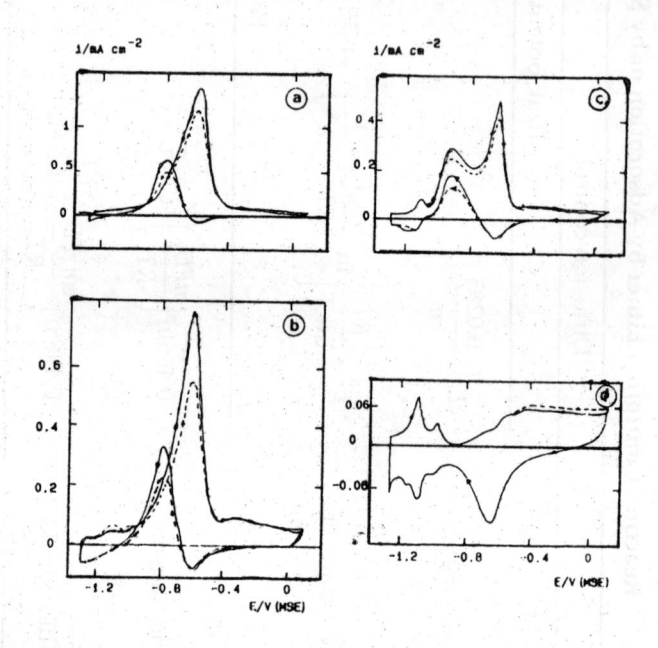

Figure 2. Cyclic voltammograms showing the oxidation of the four butanol isomers at a Pt electrode in alkaline solution (0.1 M NaOH 1 0.1 M butanol; 25°; 50 mVs⁻¹). (a) n-Butanol; (b) isobutanol; (c) butanol; (d) t-butanol.

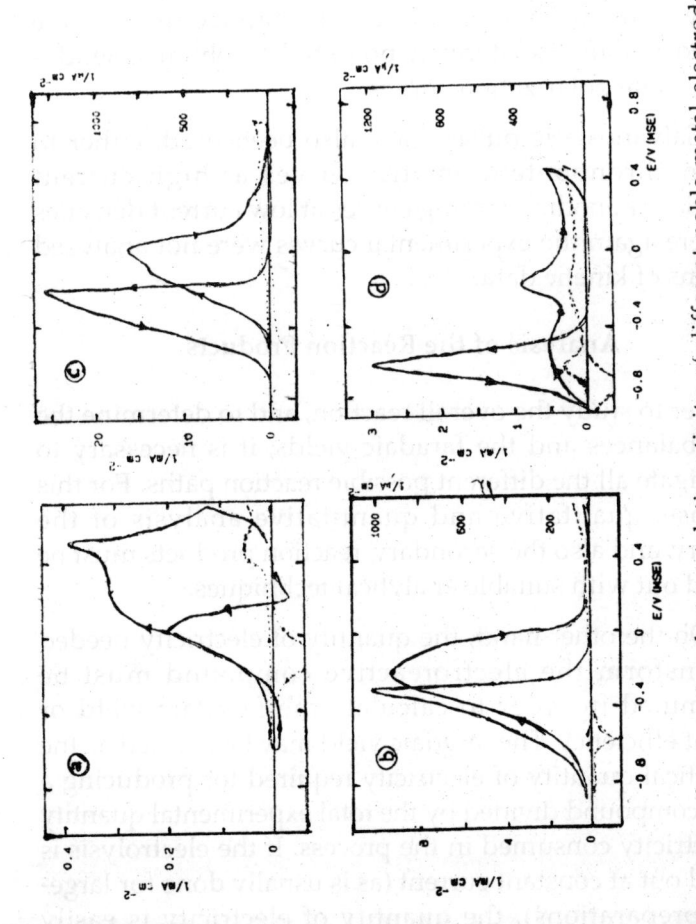

Figure 3. Voltammograms showing the electrocatalytic activity of different noble metal electrodes toward the electrooxidation of formic acid in neutral medium (0.25 M K$_2$SO$_4$ + 0.1 M HCOONa, 25°C, 50 mVs^{-1}). The dashed line in each voltammogram represents the noble metal in the supporting electrolyte. (a) Au; (b) Pd; (c) Pt; (d) Rh.

The electrooxidation of CO, formic acid and methanol has been studied by potential-step chronoamperometry. However, the $I(t)$ decay curves were not very often analyzed in terms of kinetic data. Instead, the pseudo-steady-state currents, measured at arbitrary times, were plotted versus the stepped potential to obtain pseudo-steady-state polarization curves.

Galvanostatic pulses have also been used, either to obtain chronopotentiometric curves at high current densities or anodic charging curves at low current densities but here again the experimental curves were not analyzed in terms of kinetic data.

Analysis of the Reaction Products

In order to study the overall reaction, and to determine the mass balances and the faradaic yields, it is necessary to investigate all the different possible reaction paths. For this purpose, qualitative and quantitative analysis of the primary, and also the secondary, reaction products must be carried out with suitable analytical techniques.

On the other hand, the quantity of electricity needed to transform the electroreactive compound must be determined in order to calculate the faradaic yield or current efficiency. The faradaic yield may be defined as the theoretical quantity of electricity required for producing a given compound divided by the total experimental quantity of electricity consumed in the process. If the electrolysis is carried out at constant current (as is usually done for large-scale preparations), the quantity of electricity is easily calculated, knowing the duration of the electrolysis. However, in this case, the electrochemical reaction is not controlled, since the electrode potential shifts continuously during the electrolysis, as the consequence of the disappearance of the electrolyzed compound, and since undesired electrochemical reactions may occur. Therefore, it is better to carry out the electrolysis at constant potential, particularly on the laboratory scale, in order to make the electrochemical process more selective. In that case, the quantity of electricity consumed during the electrolysis reaction is better determined using an analog or digital coulometer.

For most electrocatalytic reactions involving organic species with more than one carbon atom, the number of reaction products is relatively high. Moreover, some of the secondary products appear only in trace amounts (with concentrations smaller than 10^{-4} M). Furthermore, the different reaction products are very often similar molecules, so that some of their physicochemical properties are not very different, which complicates their detection and separation. Thus, the electroanalytical techniques used must be able to analyze a mixture of similar organic compounds at low concentrations ranging between 10^{-6} M and a few molar) dissolved in an aqueous electrolytic solution containing a concentrated supporting electrolyte.

Among quantitative electroanalytical techniques, polarography is one of the most widely used. The current height of the polarographic wave is directly proportional to the concentration of the electroactive species. However, the~ half-wave potentials of the waves, which are characteristic of the electrochemical system, are generally not very well separated for similar organic compounds, so that polarography may not always be adequate for the analysis of the products and by-products resulting from the electrocatalytic oxidation of organic compounds.

Analytical methods employed in organic chemistry are much more adaptable to the quantitative analysis of electrolysis reaction products. Both chromatographic and spectroscopic methods will be discussed in this section, particularly those allowing "on-line" analysis.

(i) Chromatographic Techniques

Chromatography is an analytical method in which the components of a mixture in a mobile phase, gaseous or liquid, are separated on an immobile phase, called the stationary phase. The stationary phase may be a solid, or a liquid supported on a solid, or a gel, either packed in a column or spread on a supporting plate as a layer or distributed as a film. The solvent of a liquid mobile phase is usually called the "eluent," whereas the inert gas of a gaseous mobile phase is called "carrier gas."

After separation of the mixture, either on a plate or through a column depending on the particular technique used, the different components are located as spots on the plate or as concentration peaks at the output of a detector. The detector output signal, plotted against time of analysis, or against the volume of the mobile phase, is called a "chromatogram" (Fig. 4).

A concentration peak in a chromatogram is characterized by its position (relative retention time, t_r, or relative retention volume, V_r), referenced to the position of a given compound which does not interact with the stationary phase, and its intensity (peak height h and width at half-height $w/2$). The retention time depends on many factors, such as the distribution coefficient of the components between the stationary phase and the mobile phase, the geometric characteristics of the column, the temperature, the pressure, and the flow rate of the eluent or of the carrier gas. Reference compounds, whose chromatograms are recorded under exactly the same experimental conditions, are used to assign a chromatogram peak to a given substance. The amount (mass or concentration) of the analyzed products is determined by comparing the integrated intensity of the chromatogram peaks (the surface area under the peak, determined by

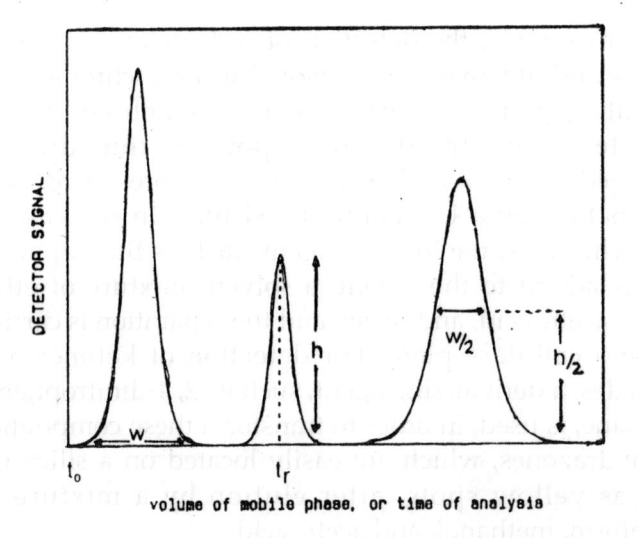

Figure 4. Theoretical chromatogram.

integration of the detector signal) to those of the reference samples at different known concentrations.

Many different types of chromatographic techniques are available, differing in the method of separation and/or the nature of the phases involved. The most widely used are described below.

(a) Thin-layer chromatography (TLC)

In thin-layer chromatography a solution of the sample in a volatile solvent is deposited at the bottom of a uniform layer of an inert adsorbent, such as silica or alumina, spread over a supporting plate made of glass or plastic. Then the plate is placed vertically in a container, the bottom of which is filled with the mobile phase. The solvent rises by capillarity and separates the sample mixture into discrete spots. At the end of the experiment, the solvent is allowed to evaporate, and the separated spots are identified either by physical methods or by chemical reactions. A reference compound is usually used for comparison.

For example, the detection of carboxylic acids, or of ketones and aldehydes, is achieved using a chromogenic reagent, called a "locating agent," which colors the separated spots (the different spots are thus directly visualized, or more easily detected by fluorescence, after illumination by a UV lamp at 254 nm). In the case of carboxylic acids, the locating agent, such as Bromophenol Blue, is added to the eluent, a solvent mixture of ethyl acetate, acetic acid, and water, and the separation is carried out on a cellulose plate. For detection of ketones and aldehydes, a derivatizing agent, such its 2, 4-dinitrophenyl hydrazine, is used, in order to transform these compounds into hydrazones, which are easily located on a silica gel plate as yellow spots, after elution by a mixture of chloroform, methanol, and acetic acid.

(b) Gas chromatography (GC)

In gas chromatography, the separation of the sample mixture, transported by an inert carrier gas (e.g., N_2, H_2, He), is achieved in a column containing the stationary phase. In gas-solid chromatography (GSC or GC), the stationary phase is a solid adsorbent, which retains selectively the different molecules of the sample (adsorption chromatography), whereas in gas-liquid chromatography (GLC or GC) the stationary phase is a liquid, which dissolves selectively the different compounds contained in the sample (partition chromatography). Separated products are characterized using various detectors, such as the flame ionization detector (FID), the thermal conductivity detector (TCD), the electron capture detector (ECD), and the nitrogen phosphorus detector (NPD); among these, the FID is the most widely used because of its reliability, high sensitivity, and good linearity.

A schematic diagram of a gas chromatograph is given in Fig. 5a.

Gas chromatography has been used to identify gaseous reaction products during the electrooxidation of methanol, methylal, and glyoxal.

(c) Liquid chromatography (LC)

In liquid chromatography, the mobile phase is a moving liquid, called the "eluent." A packed column is used as the stationary phase. This column contains either a solid adsorbent (adsorption chromatography), an ion-exchange resin (ion-exchange chromatography), a porous gel matrix (gel permeation chromatography), or a support for a liquid stationary phase (partition chromatography). High-pressure liquid chromatography, or high-performance liquid chromatography (HPLC), makes use of microparticulate materials in the column packing, which necessitates the use of high pressure pumps to overcome the large pressure drops across the column. Various detectors are now commonly used, such as the refractive index detector (RD), which detects the passage of the solute

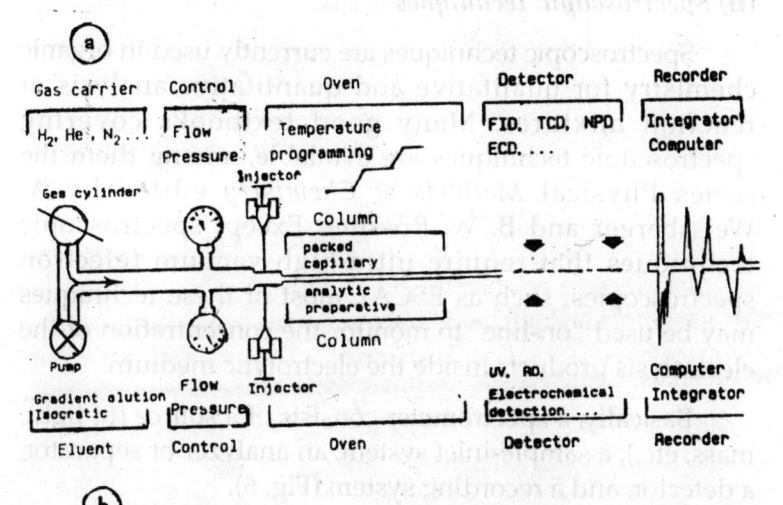

Figure 5. Schematic diagram of a gas chromatograph (a) and a liquid chromatograph (b).

by recording the corresponding refractive index change of the eluent, or a UV spectrophotometer (working either at a fixed wavelength or at variable wavelengths), which monitors the change in absorbance of the eluent as the solute passes through the detector flow cell.

A schematic diagram of a liquid chromatograph, which is similar to that of a gas chromatograph, is shown in Fig. 5b.

The development of automated liquid chromatographs has made this technique a very useful analytical tool for the qualitative and quantitative analysis of mixtures of organic compounds (even at concentrations as low as 10-6 M) in electrolyzed aqueous solution. Liquid chromatography has been successfully applied to the quantitative analysis of liquid reaction products of many electrochemical reactions, for example, reduction of glucose and oxidation of methanol, glyoxal, and 1, 2-propanediol.

(ii) Spectroscopic Techniques

Spectroscopic techniques are currently used in organic chemistry for qualitative and quantitative analysis of reaction mixtures. Many good textbooks covering spectroscopic techniques are available, among them the series Physical *Methods of Chemistry* edited by A. Weissberger and B. W. Rossiter. Except spectroscopic techniques that require ultra-high vacuum (electron spectroscopies, such as ESCA), most of these techniques may be used "on-line" to monitor the concentration of the electrolysis products inside the electrolytic medium.

Basically, a spectrometer consists of a source (of light, mass, etc.), a sample-inlet system, an analyzer or separator, a detector, and a recording system (Fig. 6).

The principles of classical spectroscopies (optical and magnetic spectroscopies) will be discussed in this section, together with those of mass spectrometry and radiometric techniques. Further details may be found in classical textbooks, whereas the use of spectroscopic methods for investigating adsorbed intermediates will be discussed in detail in Section 11.3(iv).

(a) Optical spectroscopies

Optical spectroscopies, that is, spectroscopic technique, involving the absorption by molecules of light in the UV-visible and infrared ranges, are particularly useful for the qualitative and quantitative analysis of reaction products. They are based on the dipolar interaction between the electric field vector $E(v)$ of' the incident light (electromagnetic waves characterized by their electric and magnetic field vectors, oscillating at frequency v, perpendicular to each other and to the direction of propagation) and the electrical dipole moment **p** of the organic molecule:

The electrical dipole moment **p** = q**r** (where q is the electric charge, and **r** is the vector linking the two charges of opposite signs) results either from the electron motion inside the molecule (electronic spectroscopy in the UV-

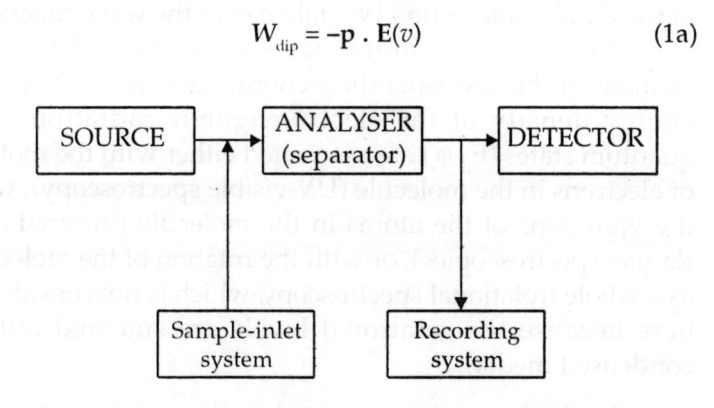

$$W_{dip} = -\mathbf{p} \cdot E(v) \tag{1a}$$

```
┌─────────┐    ┌──────────┐    ┌──────────┐
│ SOURCE  │───▶│ ANALYSER │───▶│ DETECTOR │
│         │    │(separator)│    │          │
└─────────┘    └──────────┘    └──────────┘
                     │               │
                     ▼               ▼
               ┌──────────┐    ┌──────────┐
               │Sample-inlet│  │ Recording│
               │  system   │   │  system  │
               └──────────┘    └──────────┘
```

Figure 6. Schematic diagram of a spectroscopic experiment.

visible range) or from the asymmetric distribution of the electronic charge in the chemical bonds (vibrational spectroscopy in the infrared range and Raman spectroscopy).

Absorption spectra are recorded as the light intensity at the output of the spectrophotometer versus the frequency ν, (or versus the wavenumber $\bar{\nu} = 1/\lambda = \nu/c$, or versus the wavelength $\lambda = c/\nu$, where c is the speed of light in vacuum) of the incident light. The spectra obtained exhibit absorption bands at frequencies ν_0 characteristic of the sample analyzed and related to the difference in energy between the ground state E_1 and the excited state E_2 of the molecule (Planck's relation):

$$h\nu_0 = E_2 - E_1 \qquad (2)$$

The intensity of the spectrum is proportional to the transition probability, P_{12}, between these two quantum states under the effect of the incident light:

$$P_{12} = \frac{1}{\hbar^2} |W_{12}|^2 \rho(\nu) \qquad (3)$$

where $\hbar = h/2\pi$, $W_{12} = \int \varphi_1 W_{dip} \varphi_2 \, d\tau$ is the matrix element of the dipolar interaction W_{dip} taken over the wave functions φ_1 and φ_2 of the quantum states, $d\tau$ is a small volume element of the corresponding coordinates, and $\rho(\nu)$ is the energy density of the electromagnetic radiation. The quantum states (E_i, φ_i) are associated either with the motion of electrons in the molecule (UV-visible spectroscopy), with the vibrations of the atoms in the molecule (infrared and Raman spectroscopies), or with the rotation of the molecule as a whole (rotational spectroscopy, which is not considered here because the rotational levels are smeared out in condensed media).

Qualitative analysis is based on the assignment of the characteristic frequencies to some functional groups of the

Table 2
Characteristic Frequencies of Electronic Transitiions

Group or compound	λ_{max} (Å)	ε_{max} (1 mole^{-1} cm^{-1})
C—C	1350	
C=C	1900	10,000
C=O	1900	1,000
	2800	20
O—H	1850	200
C$_6$H$_5$ (phenyl)	1950	8,000
	2500	200
H$_2$O	1667	1,480
MeOH	1835	150
Me$_2$O	1838	2,520
Benzene	2550	200

molecules (fingerprints; see Tables 2 and 3). Quantitative analysis makes use of' the Beer- Lambert law:

$$I = I^0 \, e^{-\alpha(v)Cl} \qquad (4)$$

where I^0 and I are the intensity of incident and transmitted light, respectively, l is the thickness of the absorbing layer, C is the molar concentration of the absorbing species, and $\alpha(v)$ is the linear absorption coefficient at a given frequency. This law is usualIN written as a semilogarithmic law, which allows us to define the absorbance A of the sample as

$$A = \log_{10}(I^0/I) = \varepsilon Cl \qquad (5)$$

where $\varepsilon = \alpha/2.3$ is the extinction coefficient.

A plot of absorbance versus the frequency v of the incident light, or the wavelength $\lambda = c/v$, is an absorbance spectrum.

The unknown concentration C of a given absorbing species can be obtained directly from the measurement of the absorbance if the extinction coefficient ε is known at the frequency of the measurement. Otherwise, a calibration curve is first plotted from the absorbance data for different known concentrations of the species in the sample cell.

Absorbance spectroscopy in the UV-visible range is used much more for quantitative analysis than absorbance spectroscopy in the infrared range. However the latter technique is very useful for qualitative analysis, because

Table 3
Characteristic Wavenumbers (v) for Vibrational Transitions of Diatomic Molecules and Wavenumber Ranges for Vibrational Modes

	$\bar{v}$ (cm^{-1})
Diatomic molecules	
HF	3958
HCl	2885
HBr	2559
HI	2230
HD	3817
CO	2143
NO	1876
Vibrational modes	
C—H stretching	2650-2960
C—H bending	1340-1465
C—C stretching	700-1250
C=C stretching	1620-1680
C≡C stretching	2100-2260
O—H stretching	3590-3650
C=O stretching	1640-1780
CO_3^{2-}	1410-1450
H—bonds	3200-3570
C—F stretching	1000-1400
C—Cl stretching	600-800
C—Br stretching	500-600
C—I stretching	500

infrared bands are narrower in(] are thus much more easily assigned to particular functional groups (fingerprinting) than UV-visible bands.

There are only a few papers on the analysis of electrolysis reaction products by UV-visible spectroscopy. One may cite the study of aldehyde oxidation at glassy carbon, mercury, copper, silver, gold, and nickel anodes and the study of the reduction of substituted glyoxal at a mercury electrode.

Similarly, infrared spectroscopy has been used to characterize the reaction products resulting from the oxidation of alcohols at a nickel anode in alkaline t-butanol/water mixtures.

The development of Fourier transform infrared (FTIR) spec-troscopy, associated with large computational facilities, makes infrared spectroscopy a very powerful technique for qualitative analysis of reaction products inside the electrolysis cell. The availability of large computer memories leads to a refinement of the analysis whereby a stored reference spectrum is subtracted from the sample spectrum recorded either at a given potential (SNIFTIRS) or during a potential sweep (SPAIRS); see Section II.3(iv.e). This has allowed qualitative and quantitative analysis, not only of bulk species in the reaction layer adjacent to the electrode surface, but also of adsorbed species at the electrode surface.

(b) Magnetic spectroscopies

Magnetic spectroscopies are based on the dipolar interaction between the magnetic field vector $\mathbf{B}(v)$ of the incident electromagnetic wave at a frequency v and the magnetic dipole moment μ of the molecule:

$$W_{mag} = -\mu \cdot \mathbf{B}(v) \qquad (1b)$$

The magnetic dipole moment µ is usually proportional to the spin angular momentum **I**:

$$\mu = \gamma I$$

with γ the gyromagnetic ratio ($\gamma = q/2m$). **I** is quantized; that is, its projection I_z on the z axis takes $2I + 1$ values, where I is the spin quantum number.

If the dipole moment of the molecule arises from the nuclear spin **I** of the atoms inside the molecule, the method is called nuclear magnetic resonance (NMR). If the dipole moment is associated with the electronic motion, particularly with the electronic spin angular momentum **S** (free radicals), the method is called electron spin resonance (ESR) or electron paramagnetic resonance (EPR).

Because the nuclear (or electron) spins are randomly & tribute(] inside the sample, there is no net magnetization vector **M** in the absence of a magnetic field. Moreover, the magnetic energy levels are degenerate, with a degeneracy factor equal to $2I + 1$ for $2S + 1$ for electron spins), where the spin quantum number I (or S for electrons) is integer or half-integcr.

To observe spectral transitions, it is thus necessary to apply a static magnetic field $\mathbf{B}_0$, which will induce a net magnetization vector $\mathbf{M}_0 = X\mathbf{B}_0$ (where $X > 0$ is the paramagnetic susceptibility because the energy levels are now separated into $2I + 1$ (or $2S + 1$) magnetic levels, whose population follows the Boltzmann distribu tion law.

For the transitions between the magnetic energy levels, Planck's relation [Eq. (2)] still applies, leading to the resonance equation:

$$h\nu_0 = E_2 - E_1 = g\beta B^0 \qquad (6)$$

with β the Bohr magneton ($\beta_{e'} = e_0 \hbar / 2m_e$ for the electron, $\beta_N = e_0 \hbar / 2M_{p'}$ for the proton), and g the "g factor" (the nuclear "g factor," $g_{N'}$ is related to the gyromagnetic ratio γ_N by $\gamma_N h = g_N \beta_N$).

It follows that magnetic resonance spectra are usually recorded as the absorption intensity versus the strength of the applied static magnetic field B at constant frequency v_0 of the electromagnetic wave. In the different types of magnetic spectroscopy, the frequency ranges considered are quite different. For NMR spectroscopy, radio frequency waves of a few tens to a few hundreds of megahertz (corresponding to wavelengths of a few meters) are used depending on the nuclear spin investigated (60 MHz for protons, 15 MHz for ^{13}C nuclei at a magnetic field of about 15 kT), whereas for ESR spectroscopy, microwaves of 10 GHz, or 3-cm wavelength, are used (for an X-band ESR spectrometer).

The absorption intensity is proportional to the number of spins per unit volume, thus allowing a quantitative determination of the unknown concentration of a sample. Absolute concentration measurements are usually very difficult, so that a reference sample is often used [e.g., a,a'-diphenyl-b-picrylhydrazyl (DPPH) in ESR spectroscopy].

Apart from the main interaction with the applied magnetic fields [static field B_0 and electromagnetic field $B(v)$], there are small dipolar interactions inside the molecule, leading to small variations in the positions of the energy levels. This leads to small displacements of' the positions of' the absorption bands ("chemical shift" in NMR spectroscopy, ligand field interaction in ESR spectroscopy) and to the occurrence of several lines in the spectrum (the so-called line structure in NMR due to dipolar coupling between different nuclear spins and the hyperline splitting in ESR, which arises from dipolar coupling between nuclear and electronic spins). The fine and hyperfine structures are characteristic of the molecules and, in most cases, allow their identification (in a similar way to fingerprinting in optical spectroscopies).

NMR spectroscopy has been successfully used for qualitative and quantitative analysis of electrolyzed solutions, but very few results are given in the literature. One may also cite the analysis of reaction products during prolonged electrolysis of glycerol.

ESR spectroscopy, which is a very sensitive method (as low as 10^{12} spins, i.e., about 10^{-12} mol, can be detected), has been mainly applied to the investigation of radical intermediates, which occur in the reaction mechanisms of many electroorganic reactions. Specially designed electrochemical flow cells, small enough to fit the resonant cavity, have been used so far. Most applications have dealt with organic radicals and are beyond the scope of this chapter. In the case of the oxidation of methanol and formaldehyde at a platinum electrode, simultaneous electrochemical electron spin resonance spectroscopy (SEERS) has allowed CHO radicals to be identified as reaction intermediates but because of the way in which the experiment was done, using spin traps, its relevance to electrocatalysis under normal experimental conditions is questionable.

ESR spectroscopy can also help in the understanding of electron exchange inside an electrode material, such as intercalation compounds.

(c) Other spectroscopic and related techniques

Mass spectroscopy. In classical mass spectroscopy, the organic molecule is ionized by electron bombardment, and then the molecular ion beam is accelerated by an electric field and focused on the detector, by a uniform magnetic field (mass analyzer) with a deflection depending on the ionic mass. The principle of a mass analyzer based on the deflection of the ion beam in a magnetic field is the following. When an ionic species (mass m, charge q) is accelerated by a potential difference V, it acquires a kinetic energy

$$\frac{1}{2}mv^2 = qV$$

that is, a velocity

$$v = \left(2V\frac{q}{m}\right)^{1/2}$$

When the ions enter a uniform magnetic field of strength B_0, acting perpendicular to the initial direction of motion, they experience the Laplace force $\mathbf{f} = q\mathbf{v} \times \mathbf{B}$, so that their motion becomes semicircular. The radius of the circle R is obtained by equating the magnetic force strength, $f = qvB$, acting centripetally, to the centrifugal force $f = mv^2/R$, so that

$$R = \frac{m}{q}\frac{v}{B} = \left(\frac{m}{q}\right)^{1/2}\left(\frac{2V}{B^2}\right)^{1/2}$$

or

$$\frac{m}{q} = \frac{B^2 R^2}{2V} \qquad (7)$$

For a given magnetic geometry (R fixed), the mass-to-charge ratio is thus related to the magnetic field strength B and to the accelerating potential V, so that the entire mass spectrum can be obtained, in principle, by scanning either B or V Practically, V cannot be varied by more than a factor of 10 (e.g., from 4000 to 400 V), so that magnetic field scanning i, much more advantageous for obtaining the entire mass spectrum (for $V = 4000$ V and $R = 50$ cm, a mass spectrum from 20 to 1000 m/q units is obtained by scanning the magnetic field from 8.2×10^{-2} T to 0.407 T). Other types of mass analyzer are also used, such as the quadrupole mass filter, the ion trap detector, and the time-of-flight analyzer.

Several types of detectors have been considered, such as the Faraday cup, the electron multiplier, the electro-optical ion detector, and a photographic plate. The electron multiplier detector is the most widely used, because of its excellent sensitivity (allowing the analysis of a few nanograms) and its good response time.

The introduction of the sample molecules may be either direct or through a gas-chromatographic inlet. The advantage of the latter is that the sample mixture is separated before its introduction into the ionization chamber of the mass spectrometer. The gas chromatography/ mass spectroscopy (GC-MS) combination is a very powerful technique for the analysis of mixtures, which, without chromatographic separation, would lead to extremely complicated mass spectra (with many peaks for each component) that would be impossible to interpret correctly.

The ionization of the molecule is usually achieved by interaction with accelerated electrons (typically having an energy of 70 el emitted from a hot filament. Electron bombardment of a molecule AB leads to several ions, including parent ions (AB^+, AB^{2+}, or AB^-) produced by ionization of the molecule, fragment ion (A^+, B^+, A^-, B^-, etc.) produced by dissociation of the molecule and rearrangement ions produced by redistribution of atoms or groups of atoms. Thus, a given molecule will give rise to a large number of peaks in the mass spectrum (fingerprint analysis).

Other modes of ionization, such as field ionization, which employs an electric field of 10^7 to 10^8 V cm^{-1}, transfer a much smaller energy to the molecule, so that predominantly parent ions are produced, making the spectra easier to interpret.

On-line mass spectrometry was first used in electrochemistry by Bruckenstein and Comeau, to monitor the gas evolved during electrochemical reactions. The method was further developed by Heitbaum, Vielstich, and co-workers to investigate reaction mechanisms of the electrocatalytic oxidation of small organic molecules (methanol, formic acid, carbon monoxide). In these experiments, the cyclic voltammogram and the mass intensity versus potential curves of the gaseous and volatile species involved were 114 recorded simultaneously.

The nature of the primary reaction products of the electrocatalytic oxidation of glucose in phosphate buffer solutions has been determined by "on-line" mass spectroscopy. Adsorbed intermediates produced during the chemisorption of small organic molecules (e.g., HCOOH, CH_3OH, C_2H_5OH) have also been investigated by differential electrochemical mass spectroscopy (DEMS), by electrochemical thermal desorption mass spectroscopy (ECTDMS), and by secondary ion mass spectroscopy (SIMS). More details will be given in Section II.3(v).

Tracer methods. Tracer methods consist in introducing into, the electrochemical system trace amounts of an isotope of a particular element (tracer) in the same chemical form as the investigated compound and following its fate during the electrochemical reaction. The detection of the isotopically labeled compound is achieved either by mass spectroscopy (tracer method) for stable isotopes or by radiometry (radioactive tracer method) for radioactive isotopes.

Although tracer methods are available for qualitative analysis of reaction products, particularly after their separation by chromatographic techniques," they have been mainly used in electrochemistry for adsorption studies, as discussed in Section II.3(vi).

5

Study of the Adsorbed Species

(i) Introduction

In electrocatalytic reactions, as in heterogeneous, catalysis, the surface species, particularly the adsorbed species, play a key role in the activity of the catalytic electrode (as reactive intermediates, poisoning species, etc.) and in the selectivity of the reaction (as primary determinant of the elementary steps and the pathways taken in complex reactions).

Let us consider an electrocatalytic reaction involving an oxygenated organic compound, $C_xH_yO_z$. Its complete oxidation to carbon dioxide may require the donation of oxygen atoms, which come from water molecules, or water ions, in aqueous medium:

$$C_xH_yO_z + pH_2O \rightarrow xCO_2 + nH^+ + ne^-$$

with $p = 2x - z$ and $n = 4x + y - 2z$.

In fact, even for the oxidation of a small molecule, such as methanol, the overall reaction is relatively complex and involves several electron transfers; for example, in acid medium,

$$CH_3OH + H_2O \rightarrow CO_2 + 6H^+ + 6e^-$$

or, in alkaline medium,

Study of the Adsorb Species

$$CH_3OH + 8OH^- \rightarrow CO_3^{2-} + 6H_2O + 6e^-$$

Such complex reactions proceed through multiple elementary steps and pathways, each of them involving no more than one or two electrons. The interaction of the catalytic surface with the electroreactive species leads to the formation of different adsorbed intermediates, some of them acting as reactive intermediates (RI), others as poisoning intermediates (PI); see Section III.1.

This is illustrated by the oxidation of formic acid on platinum electrodes, which is sometimes considered as one of the most simple electrocatalytic reactions. At potentials greater than 0.45 V versus RHE, where adsorbed hydrogen is ionized, this reaction can be written as follows:

$$HCOOH \begin{array}{l} \nearrow -COOH_{ads} \text{ (RI)} + H^+_{aq} + e^- \\ \searrow -CO_{ads}\text{(PI)} + H_2O \end{array}$$

$$-COOH_{ads} \text{\textcircled{R}} CO_2 + H^+_{aq} + e^-$$

$$H_2O \text{\textcircled{R}} -OH_{ads} + H^+_{aq} + e^-$$

$$CO_{ads} + OH_{ads} \text{\textcircled{R}} CO_2 + H^+_{aq} + e$$

The poisoning intermediates are strongly bonded to the electrode surface, thus blocking the electrode active sites and causing the electrode activity to decrease with time. The role of poisoning intermediates in the electrooxidation of most organic molecules has been recognized for a long time. They were called O-type species by Breiter, whereas the weakly bonded desorbable species were called C-H-type species. The strongly bonded species only oxidize at potentials greater than 0.6 V versus RHE, potential at which the platinum surface begins to oxidize, and behave similarly to chemisorbed carbon monoxide [see Section III.1(iii)].

(ii) Adsorption Isotherms and Adsorption Kinetics

The reaction kinetics depend strongly on the amount of adsorbed species, that is, on the degree of coverage θ_i of the electrode surface by adsorbed species i.

The electrode surface coverage θ_i may be defined as the ratio of the number of adsorbed species N_{ads} per unit area S (surface concentration $\Gamma = N_{ads}/S$) to the maximum surface concentration Γ^m:

$$\theta = \Gamma/\Gamma^m \qquad 0 \leq \theta \leq 1 \qquad (8a)$$

It can also be defined as the ratio of the number of adsorbed species, N_{ads}, to the number of adsorption sites on the substrate, N_s:

$$\theta = N_{ads}/N_S \qquad (8b)$$

However, in the latter definition, there remains some ambiguity, because the number of adsorption sites on the Substrate may depend on the nature of the adsorbed molecule and on the heterogeneity of the surface. Moreover, multilayer adsorption may occur so that θ may be greater than unity. So, another alternative is to define θ as the fraction of the surface area occupied by the adsorbed species ($0 \leq \theta \leq 1$). This implies that the total surface area which can be blocked by adsorbed organic species is known.

The determination of the real active surface area S is a very important point, since the current density $i = I/S$, where I is the measured current intensity, has to be calculated in order to compare the electrocatalytic activity of different electrodes. As discussed in Section 11.4(ii.a), S can be easily determined for catalytic electrodes (e.g., Pt, Rh) from the quantity of electricity Q_H^o required to deposit or to oxidize a full monolayer of adsorbed hydrogen, provided that the theoretical quantity of electricity, Q_H^{theor}, required per unit surface area (1 cm² usually) is known.

If Q_H is the charge required to deposit hydrogen on the adsorption sites free of organic adsorption (free surface), θ can be written as

$$\theta = 1 - \frac{Q_H}{Q_H^o} = \frac{Q_H^o - Q_H}{Q_H^o} \tag{9}$$

Thus, this definition implicitly assumes that the adsorption sites are the same for hydrogen and for the organic molecule.

One may determine the number of adsorbed organic species from the knowledge of the quantity of electricity Q_{org} necessary to oxidize them completely to CO_2; that is,

$$Q_{org} = n e_0 N_{ads}$$

with n the number of electrons involved (number of electrons per molecule or N_{epm}), and e_0 the elementary charge.

Thus,

$$\theta = \frac{Q_{org}}{Q_{org}^m} \tag{8c}$$

where Q_{org}^m, is the maximum value of Q_{org}, achievable at saturation coverage of the surface by the organic species. The number of electrons per molecule can be calculated as

$$N_{epm} = \frac{Q_{org}}{e_0 N_{ads}} \tag{10}$$

provided that N_{ads} can be determined independently (e.g., by radiotracer methods). Otherwise, some hypothesis regarding the value of N_{epm} (which must be an integer) can be made.

The number of electrons per site (N_{eps}) is then defined as the number n of electrons required to oxidize the adsorbed species divided by the number x of hydrogen sites that the adsorbed species occupies:

$$N_{eps} = \frac{n}{x} = \frac{Q_{org}}{e_0 N_{ads}} \bigg/ \left(\frac{Q_H^0 - Q_H}{e_0 N_{ads}} \right)$$

Therefore,

$$N_{eps} = \frac{Q_{org}}{Q_H^0 - Q_H} \qquad (11)$$

The number of electrons per site is easily measurable since one needs only to determine the quantity of electricity required either to completely oxidize the adsorbed residue (Q_{org}) or to deposit or remove adsorbed hydrogen (Q_H^0 and Q_H).

The determination of N_{eps}, which is a ratio of two integers, n and x, may help in the elucidation of the nature of the chemisorbed species X_{ads} arising from the adsorption of C_1 molecules, as shown in Table 4. However, the assumptions are not at all unambiguous, as discussed in Section III.1, since the same N_{eps} (e.g., 1) holds for different adsorbed species, and since a mixture of different adsorbed species may lead to an averaged intermediate value of N_{eps}.

In the electrocatalytic oxidation of small organic molecules, the chemisorption process is usually dissociative, that is, quite irreversible. However, it is usual to relate θ_i to the concentration C_i of a reactive species *i* in solution and to describe adsorption isotherms in the same way as at the gas/solid interface. Details on the derivation of isotherms and on the different types of isotherms can be

Table 4
N_{eps} for Adsorbed C_1 Molecules

X_{ads}	N_{eps}
– CHO	3
– CO \\	2
CO / \\	1
– COH /	1
– COOH	1
\\ – CO /	0.66

found in review papers. The adsorption isotherms most often encountered at the electrode/electrolyte interface are the Langmuir isotherm:

$$\frac{\theta_i}{1-\theta} = K_i C_i \quad \text{with } q = \sum_i \theta_i \quad (12a)$$

and the Temkin-Frumkin isotherm:

$$\frac{\theta_i}{1-\theta} = K_i C_i e^{-g\theta_i} \quad (12b)$$

where K_i is the adsorption equilibrium constant (related to the adsorption Gibbs free energy; $\Delta G_i^a = -RT \ln K_i$), and g is an interaction factor between adsorbed species.

The adsorption of organic species at the electrode/electrolyte interface also depends strongly on the electrode potential E and is therefore called "electrosorption." The dependence of dissociative chemisorption on E is relatively complex, since E affects the interaction of both the water adsorption residues (H_2O, H_{ads}, OH_{ads}) and the adsorbed organic species with the electrode surface and changes the amount of adsorbed species by oxidation or reduction.

Therefore, bell-shaped adsorption curves $\theta(E)$ are usually obtained for the electrosorption of organic compounds.

The kinetics of adsorption is also an important limiting factor which may play a significant role in the overall reaction kinetics. Kinetic laws may be obtained by plotting the measured concentration of adsorbed species (or a quantity, such as Q_{org}, directly related to it) as a function of adsorption time. The simplest kinetic law is obtained under Langmuir conditions. For the adsorption process of' species i, it is written as

$$v_{ads} = \frac{d\theta_i}{dt} = k_a C_i (1-\theta) \tag{13}$$

and for the desorption procees as

$$v_{des} = -\frac{d\theta_i}{dt} = k_d \theta_i \tag{14}$$

where k_a and k_d are the rate constants for adsorption and desorption, respectively.

The adsorption kinetics for small organic C_1 molecules is very often governed by the Roginskii-Zel'dovich equation:

$$v_{ads} = \frac{d\theta_i}{dt} = k_a C_i \exp(-\beta f \theta_i) \tag{15}$$

where f is the heterogeneity factor of the surface, and β is a symmetry factor (equivalent to the transfer coefficient α in electrode transfer reactions). This kinetic law will lead to the Temkin adsorption isotherm:

$$\theta_i = a + \frac{\ln C_i}{f} \tag{16}$$

which is an approximation of the Temkin-Frumkin isotherm for $0.2 \leq 5\, \theta_i \leq 0.8$. This shows that the surface heterogeneity factor, $-$, is equivalent to the interaction factor, g, between adsorbed molecules.

When the reaction rate is controlled by both adsorption and diffusion, the rate expressions are much more difficult to write. In simplified cases, such as at small adsorption times for which the concentration $C_i(0, t)$ of the electroactive species at the electrode surface remains small compared to the bulk concentration C_i^0 [$C_i(0, t) \leq C_i^0$], the adsorption kinetics under diffusion control may be written as

$$\theta_i(t) = \frac{C_i^0}{\Gamma_i^m}\left(\frac{2D_i}{\pi}\right)^{1/2} t^{1/2} \qquad (17)$$

where D_i is the diffusion coefficient of species i in solution, and Γi^m is the maximum surface concentration. This predicts a linear $\theta(t)$ versus $t^{1/2}$ law for small adsorption times.

However, for the general case, the equations are much more complicated, and the reader is invited to refer to a few specialized papers.

The elucidation of electrooxidation reaction mechanisms, and of the role of catalytic electrodes, thus requires a detailed knowledge of the nature, structure, and amount of the different adsorbed intermediates involved.

In the, next section, the experimental -methods that may be employed to determine the amount of adsorbed species, that is, the degree of coverage q_i, and the nature and structure of adsorbates will be presented. Electrochemical methods, the description of which may be found in many textbooks of electrochemistry, will be briefly presented. Then, spectroscopic methods and other physical methods which allow investigation in situ of the adsorbed species at the electrode/electrolyte interface will be discussed in more detail.

(iii) Electrochemical Methods

(a) Coulometric method

Most of the electrochemical methods used to study adsorption processes are based on the quantitative determination of the electrical charge needed to oxidize the adsorbed organic substance, Q_{org}, or to deposit an adsorbed layer of other adsorbed species, such as Q_H for hydrogen adsorption and Q_o for oxygen adsorption. The degree of coverage and the number of electrons per site are easily determined from these quantities, as discussed in Section II.3(ii) [see Eqs. (9)-(11)].

Many electrochemical methods are currently used for determining the degree of coverage θ_i of adsorbed species i; among these, cyclic voltammetry, pulse voltammetry, and multipulse potentiodynamic methods are the most popular. All of them are based on the application of a programmed sequence of potentials to the electrode surface and on the measurement of the quantity of electricity Q involved.

Discussions on the choice of experimental conditions (sweep rate, etc.) to obtain reliable results and on the different correcting factors (charge of the double-layer capacity, displacement of oxygen adsorption by organic adsorption, etc.) may be found in books and in specialized papers.

Before the development of potentiodynamic and voltammetric methods, coulometry was performed at constant current intensity I, so that the quantity of electricity Q involved in the oxidation or in the reduction of the adsorbed organic species was easily obtained as the product of I and some transition time τ corresponding to the end of the electrochemical process. The so-called "charging curves," $E = f(Q)$, were thus obtained as the variation of electrode potential E with the electrolysis time, that is, with the quantity of electricity Q. The method is equivalent to chronopotentiometry, but the variation of electrode potential is due to variation of the surface concentration of the adsorbed species, whereas in classical chronopotentiometry the variation of E with time results from a diffusion process (concentration overvoltage). The transition time T is relatively well defined, since the potential arrest during which the organic substance is oxidized (anodic charging curve) or reduced (cathodic charging curve) is followed by a rapid potential variation resulting from oxygen or hydrogen adsorption, respectively. Fast galvanostatic charging curves allow the determination of the coverage in the presence of organic molecules dissolved in the electrolyte, whereas for slow galvanostatic charging curves, it is necessary to remove the parent molecule using a flow cell or to transfer the electrode surface under inert atmosphere to a cell containing the supporting electrolyte alone. Voltammetric curves $I(E)$ (where the electrode potential is a linear or triangular function of time, with a sweep rate $v = \pm dE/dt$), are in fact equivalent to the derivative of the charging curves $E(Q)$, since

$$I(E) = \frac{dQ}{dt} = \frac{dQ}{dE}\frac{dE}{dt} = v\frac{dQ}{dE} = \frac{v}{(dE/dQ)}$$

Moreover, the resolution (that is, the ability to distinguish between different processes occurring at similar potentials) of derivative curves, such as voltammetric curves $I(E)$, is much better than that of integral curves, such as charging curves.

The quantities of electricity Q are thus easily calculated by integration of the part of the voltammograms corresponding to current peaks:

$$Q = \int_{t_1}^{t_2} I(t)dt = \int_{E_1}^{E_2} \frac{I(E)dE}{v} \tag{18}$$

where E_1 and E_2 are the potentials which define the beginning and the end of the voltammetric peakos (see Fig. 1).

(b) Differential capacity and impedance methods

The differential capacity C_d of the double layer can be measured from the potential decay curves obtained once the electrode, previously polarized at a given overvoltage η_0, has been disconnected from the external electrical current. Thus, the double layer capacity discharges into the resistance of charge transfer R_c, and the analysis of the $\eta(t)$ curve allows C_d to be determined :

$$\eta(t) = \eta_0 e^{-t/R_c C_d}$$

More generally, C_d can be determined from impedance measurements. In the simple case where it can be assumed that the electrical double-layer capacity is in parallel with the charge transfer resistance R_c (or more generally with the faradaic impedance Z_f), both of them being in series with the electrolytic resistance R_e, the following equivalent circuit (Randles equivalent circuit) applies:

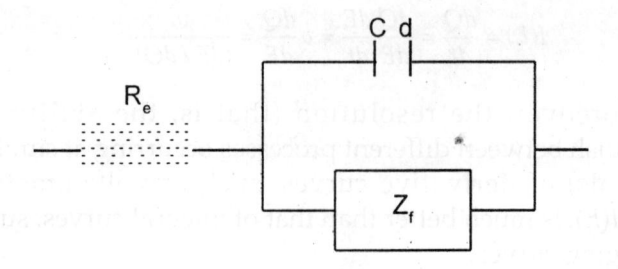

the impedance of which is given by

$$Z = \frac{dE}{dt} = R_c + \frac{Z_f}{1 + j\omega C_d Z_f} \qquad (19)$$

Study of the Adsorb Species

with ω the angular frequency of the sinusoidal voltage, and $j = \sqrt{-1}$. In the case of semi-infinite linear diffusion, Z_f is expressed by

$$Z_f = R_c + Z_w \quad (20)$$

where the Warburg impedance Z_w is

$$Z_w = (1-j)(\sigma_0 + \sigma_R)/\sqrt{\omega} \quad (21)$$

with

$$\sigma_0 + \sigma_R = \frac{(\delta I/\delta C_O)_{E,C_R}}{nFS\sqrt{2}\sqrt{D_O}(\delta I/\delta E)_{C_O,C_R}}$$

$$+ \frac{(\delta I/\delta C_R)_{E,C_O}}{nFS\sqrt{2}\sqrt{D_R}(\delta I/\delta E)_{C_O,C_R}} \quad (22)$$

The frequency analysis of the electrode impedance $Z(\omega)$ allows the different components of the Randles equivalent circuit to be determined, in particular, C_d.

The determination of the degree of coverage θ from C_d is based on the Frumkin equation, which states that the electrical charge of the metal surface is a linear function of θ; that is,

$$q(\theta) = q_0(1-\theta) + q_1\theta \quad (23)$$

where q_0 and q_1 are the charges on the bare electrode and on the completely covered electrode, respectively. Differentiating this equation with respect to potential gives

$$C_d(\theta) = \frac{dq(\theta)}{dE} = C_0(1-\theta) + C_1\theta + (q_1 - q_0)\frac{d\theta}{dE}$$

For small variations of the coverage with potential, dq/dE can be neglected, so that the degree of coverage q is given by

$$\theta = \frac{C_d(\theta) - C_0}{C_1 - C_0} \quad (24)$$

This equation applies in the range around the potential of zero charge, where the adsorption is maximum and $d\theta/dE \approx 0$.

When the adsorption kinetics is rate-limiting, the faradaic impedance Z_f is modified and contains additional capacitance and resistance terms associated with the adsorption process. This complicates greatly the impedance analysis.

(iv) Spectroelectrochemical Studies of the Adsorbed Species

Electrochemical methods are not sufficient for elucidation of the detailed nature and structure of the adsorbed species, since they are based on measurements of macroscopic quantities, proportional to the number of species involved, such as the current intensity, the quantity of electricity, or the double-layer capacity. They are not able to recognize from which molecule the electrical current, that is, the electron flow, comes, and to which species it will go. Only methods based on atomic and molecular phenomena, such as spectroscopic techniques, particularly those allowing in situ analysis of the electrode/ electrolyte interface, will be able to give information on the exact nature and structure of the adsorbed layers.

(a) In situ spectroscopic methods

Elucidation of the nature of adsorbates and their conformation on catalytic surfaces has always been a challenge.

Recent years have seen considerable progress made in the study of the solid/gas interface with the development of ex situ ultra-high-vacuum (UHV) techniques, such as electron diffraction techniques (e.g., LEED, RHEED) as well as electron spectroscopies (e.g., UPS, XPS, ESCA, AES).

Presently, similar information is needed for further development of interfacial electrochemistry, not only on the structure of the electrode surface itself, but also on the adsorbates and on the solution in the vicinity of the electrode surface.

Transferring the electrode to the ultra-high-vacuum chambers required by the above techniques is certainly an interesting idea, and various approaches have been tried by several groups (for a review). Valuable information is obtained on the structure of the electrode surface, especially with single crystals, provided that the transfer is done with great care to avoid any contamination by residual gas and provided that removal of the electrode from solution is not followed by surface restructuring during transfer to high vacuum. Obviously, only very limited information on adsorbates can be obtained from these techniques (the species have to be strongly chemisorbed), and no information at all can be expected on the structure, or the composition of the solution in contact with the electrode surface (i.e., in the double layer).

Therefore, there has been a need for the development of other types of techniques that are nondestructive with respect to the electrode/ solution interface. Since the beginning of the 1980s, successful approaches have been made in different directions. We will consider those that can combine electrochemical techniques with in situ methods, such as reflection- absorption spectroscopies, diffraction techniques, and mass spectrometry (with a special emphasis on the former, due to the considerable recent development of related techniques), separately from those based on the measurement of the amount of matter (i.e., radiometry or gravimetry with quartz microbalances).

If there is one factor common to the various in situ techniques available to investigate the electrode/ solution interface (i.e., reflection-absorption spectroscopy in the UV-visible and in the infrared range, or X-ray reflection or diffraction), it is the necessity to design a special electrochemical cell compatible with both the inescapable experimental constraints inherent to electrochemistry (particularly because of the presence of an absorbing solvent) and the geometric restrictions due to reflectance at the interface. Thus, the following criteria must be met:

(i) all the materials have to be stable in contact with the electrolytic solutions;

(ii) windows, if not the whole cell, have to be transparent to the radiation used;

(iii) if the external reflection mode is selected (Fig. 7a), the solution may drastically absorb the radiation, but the solution thickness cannot be decreased too much (typically not to less than a few microns) for the potentiostatic control of the electrode surface to operate normally; and

(iv) if the internal reflection mode is chosen (Fig. 7b), there are no restrictions on the solution thickness, but, as the transparent substrate is generally not the material that one wants to use as the electrode (and may even not be conducting), a metallic layer has to be deposited onto its surface to serve as the working electrode; the catalytic properties of this metallic layer certainly differ considerably from those of the smooth metal. Moreover, it is not easy to vary the electrode structure, particularly by using different single crystal faces.

Furthermore, in any case, the polarization of the incident beam is affected upon reflection. This leads to the so-called "surface selection rule," as first recognized by Francis and Ellison and quantitatively established by Greenler et A in the infrared range. 1 13,114

The concentration of species to be detected in studies of monolayers or submonolayers of adsorbates is always

Study of the Adsorb Species

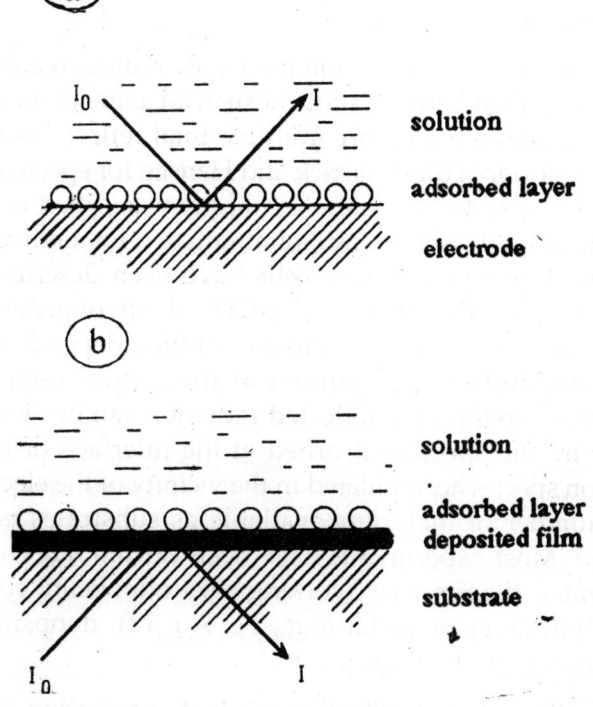

Figure 7. External (a) and internal (b) reflection modes.

low relative to that of the bulk species (the number of surface species is of the order of magnitude of the number of crystallographic sites, that is, ca. 10^{15} species per square centimeter, to be compared with 10^3 to 10^5 more species in solution), so that conventional spectroscopy is totally unable to sort out the signals due to surface species. More sophisticated techniques, using data acquisition coupled with signal processing or enhancement of the signal-to-noise ratio with phase sensitive detection, are necessary to extract the signals due to the surface species from the background (bulk species) and from the noise.

Depending on the wavelength range of the radiation and the reflection mode chosen, different types of spectroelectrochemical cells are used, and examples may be found in the literature.

In the internal reflection mode, the radiation enters an optically transparent substrate at such an angle of incidence that a single or preferably multiple total reflections occur. Initially developed by Harrick and Hansen for spectroscopy in the UV-visible range, the technique was called internal reflection spectroscopy (IRS) or attenuated total reflectance (ATR). More recently, new cells have been described for applications in the infrared range. The depth of penetration of the "evanescent wave" into the solution depends on the wavelength, but is maximum at the critical angle (see below). Therefore, the reflected radiation can be absorbed either by the species adsorbed at the interface or by the solution species accumulated in the vicinity of the electrode. The number of materials available as substrates is very limited. Most experiments have been carried out with Ge substrates, the top face (working electrode) being covered by a thin layer of metal (e.g., Fe, Au, Pt), deposited by evaporation under vacuum.

In the external reflection mode, the radiation crosses the solution, and the cell design depends on how much it is attenuated, that is, how long the path into the solution can be, the longest being the best from the electrochemistry point of view.

For spectroscopy in the UV-visible range, the solution is generally weakly absorbing, so that the simplest cell design involves sealing two quartz windows (oriented at angles close to 90^0) to a Pyrex cylinder, with the flat disk working electrode (about 1 cm^2) attached at the extremity of a syringe barrel, the trunk of the syringe being also sealed to the glass cell and oriented along the bisector of the angle formed by the axes of the two windows. The grid counter electrode is positioned away from, and parallel to, the working electrode, while the tip of a Luggin capillary (linked to the reference electrode) is placed close to the working electrode. As the beam path into the solution can be as long as a few centimeters without inconvenience and as the syringe allows the position of the working electrode to be adjusted accurately, these cells are very convenient for optical alignment. They were described originally by Bewick and Tuxford and Kotz and Kolb.

The case of external reflectance infrared spectroscopy is more complicated, in that the solution absorbs the radiation so strongly (especially with water) that a solution layer more than a few microns thick is not acceptable. In the original design by Bewick et al., the infrared radiation crosses the window at an angle of incidence of 65 to 70^0 and then passes through a thin layer of solution before being reflected by the electrode surface and then passing again through the solution and the window. A wire ring counter electrode is placed round the hollow syringe, and the tip of the Luggin capillary is positioned on the side, as close as possible to the working electrode. Although the design is far from being perfect (from the perspective of optics, the angle of incidence is too low, and from the perspective of electrochemistry the solution layer is too thin, especially if the reaction is mass-transfer limited), but it is a good compromise and the cell has the advantage of simplicity. No real improvement has been achieved despite some attempts to design different shapes of windows (especially to diminish the energy losses by reflection at the external surface of the window) or to monitor the solution thickness for optimization of the signals, but always at the expense of simplicity and rapidity of use. A limited number of infrared materials are stable in contact with acids or bases and can be used as windows. The list includes CaF_2, Si, SiO_2, and ZnSe to cover the whole range from 500 to 5000 cm^{-1}, but none of them can be assumed to be totally inert under the experimental conditions so that frequent polishing of the window is necessary.

A very similar cell was used by Fleischmann et al. for in *situ* X-ray diffraction studies. In that case the window was made of Mylar, and 100-μm-thick spacers were used to control the solution thickness.

(b) *Physical basis for reflection-absorption spectroscopy in the UV-visible and infrared ranges*

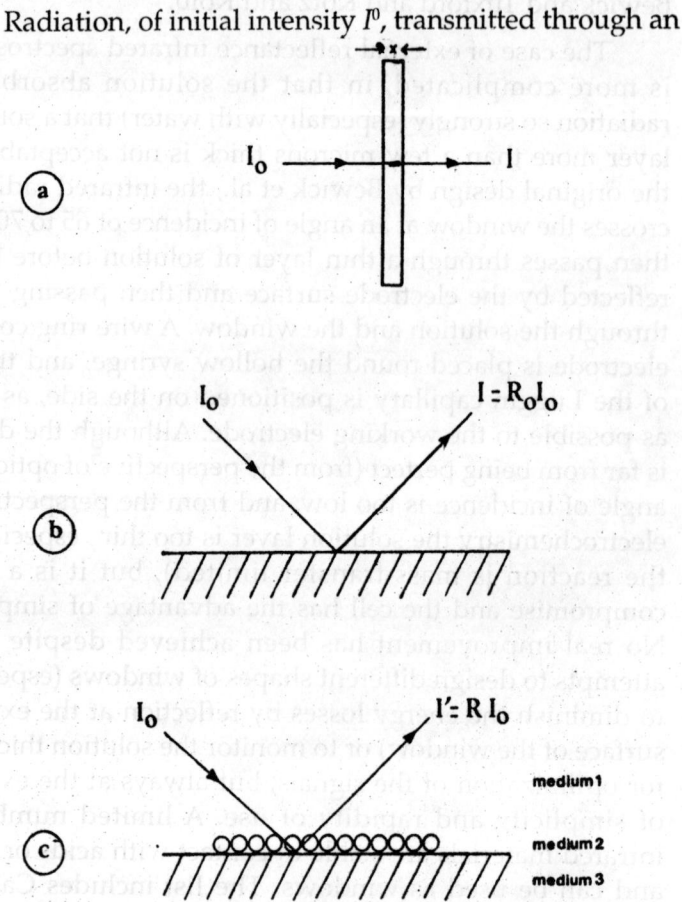

Radiation, of initial intensity I^0, transmitted through an

Figure 8. Different modes of light propagation. (a) transmission spectroscopy; (b) reflection at a nonabsorbing plane; (c) reflection-absorption spectroscopy.

Study of the Adsorb Species

absorbing layer of thickness x (Fig. 8a) emerges with intensity I, according to the Beer-Lambert law [see Eq. (4)].

Similarly, in the reflection mode, let us consider now the two limiting cases where the reflecting surface is M practically free of any absorbing species (Fig. 8b) and (ii) fully covered by a monolayer of absorbing species (Fig. 8c). Let us also assume that medium 1 is a solution, that the substrate (medium 3) is a metal (thus conductive), and that the thin film (medium 2) is isotropic. Thus, if R^0 is the reflectivity coefficient of the surface covered by the transparent layer (ideally a two-phase model), and T the reflectivity coefficient of the surface in the presence of the absorbing layer (the three-phase model), then the resulting intensities of the reflected beams can be expressed, respectively, by:

$$I = R^0 I^0 \tag{25a}$$

$$I' = R I^0 \tag{25b}$$

and the relative reflectivity change, $\delta R/R^0$, of the surface becomes

$$\delta R/R^0 = (R - R^0)/R^0 = (I' - I)/I \tag{26}$$

The variation of the absorbance, as defined in Eq. (5), due the presence of the film is then

$$\delta A = \ln(I^0/I') - \ln(I^0/I) = -\ln(I'/I) \tag{27}$$

Thus, using Eq. (26):

$$\delta A = -\ln(1 + \delta R/R^0) \approx -\delta R/R^0 \tag{28a}$$

for $\delta R/R^0 \ll 1$, which is usually the case for adsorbed films or thin layers.

δA is therefore equivalent to the absorption factor originally defined by Greenler as

$$\delta A = (R^0 - R)/R^0 \tag{28b}$$

Therefore, the absorption spectrum of an adsorbed layer can be obtained by reflectivity measurements, and the technique is called reflection- absorption spectroscopy.

According to the theory, a linearly polarized electromagnetic plane wave becomes elliptic upon reflection, the degree of ellipticity depending on the angle of incidence, that is, on how much the two components of the electric field [the p-polarized vector parallel to the incidence plane (defined by the propagation direction of the incident beam and the normal to the surface) and the s-polarized vector perpendicular to it, and thus parallel to the interface; see Fig. 9] are affected by phase changes and amplitude changes. A technique based on the discrimination of these two vectors in the infrared range was successfully developed by Stobie et al. and is called infrared ellipsometry.

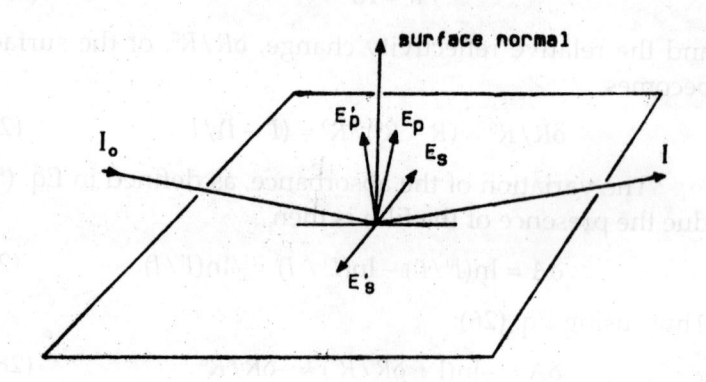

Figure 9. Changes in s- and p-polarized components of the electric field upon reflection at a high angle of incidence.

Table 5 gives the changes in phase angles upon reflection for the two s- and *p*-polarized electric field vectors, where φ_i and φ_r are the angles of incidence and refraction, respectively. The particular value φ_{iB} of the angle of incidence such that $(\varphi_{iB} + \varphi_r) = \pi/2$ is the critical angle,

Study of the Adsorb Species

or "Brewster angle," at which the reflected light becomes totally s-polarized [see Eqs. (30) below]. Using Descartes' law [see Eq. (31)], φ_{iB} is calculated from the relation tan $\varphi_{iB} = n_2/n_1$, where n_1 and n_2, the refractive indices of the two media, are taken as real (but they would become complex for highly absorbing media; see below).

Whatever the angle of incidence and the refractive indices are, the s-component of the electric field is reversed upon reflection (i.e., its phase angle is changed by π). Assuming that the reflectivity coefficients are not very different, the consequence is that the resultant of the incident and reflected s-vectors is nearly zero at the surface. The situation is different for the p-vectors. In the usual case where n_2 is greater than n_1 (this is the case for the electrochemical interface), the reflected *p-polarized component is reversed at* $\varphi_i < \varphi_{iB}$ *and not reversed at* $\varphi_i > \varphi_{iB}$. In the latter case, the incident and reflected p-polarized vectors of the electric field, at the surface, coadd in such a way that the resultant vector is enhanced. Actually, its magnitude depends strongly on the angle of incidence. It is usual to denote by $E_{p\perp}$ the projection of the resultant field vector on the normal to the reflection surface, and by $E_{p||}$, its projection on the surface (Fig. 10). It is clear from Fig. 11 that $E_{p\perp}$ can be nearly doubled at high angles of incidence, while $E_{p||}$, as well as E_s, always remains very small.

Table 5
Changes in Phase Angles upon Reflection

	Changes in phase angle for	
	$\varphi_i + \varphi_r < \pi/2$	$\varphi_i + \varphi_r > \pi/2$
p-Polarization	π	0
s-Polarization	π	π

The consequence for *reflection-absorption spectroscopy* is that the s-polarized component of the electric field is always inactive while the *p*-polarized component can interact strongly with those vibrational modes that correspond to dipoles oscillating perpendicularly to the surface. This is the origin of the so-called "*surface selection rule*" in infrared reflectance spectroscopy, which holds for Raman and UV-visible spectroscopies as well.

Complete analytical expressions for $(\delta R/R)_p$ and $(\delta R/R)_s$ were given by McIntyre and Aspnes, starting from the Fresnel equations for the reflection coefficients.

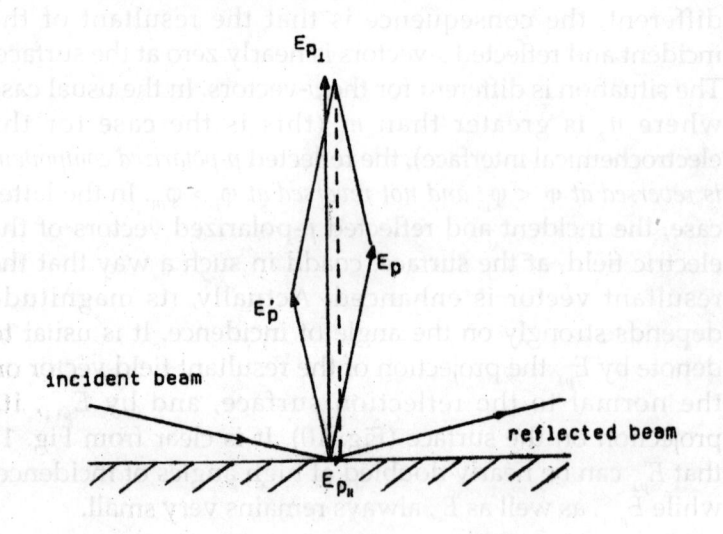

Figure 10. Normal $E_{p\wedge}$ and tangential $E_{p||}$ components of the p-polarized resultant electric field vector.

The reflectivity coefficient R of the interface is

$$R = \frac{I_r}{I_i} = \frac{\langle E_r^2 \rangle}{\langle E_i^2 \rangle} \qquad (29)$$

since the light intensity of the incident beam I_i and that of the reflected beam I_r are proportional to the mean square of the electric field vectors E_i and E_r, respectively.

Study of the Adsorb Species

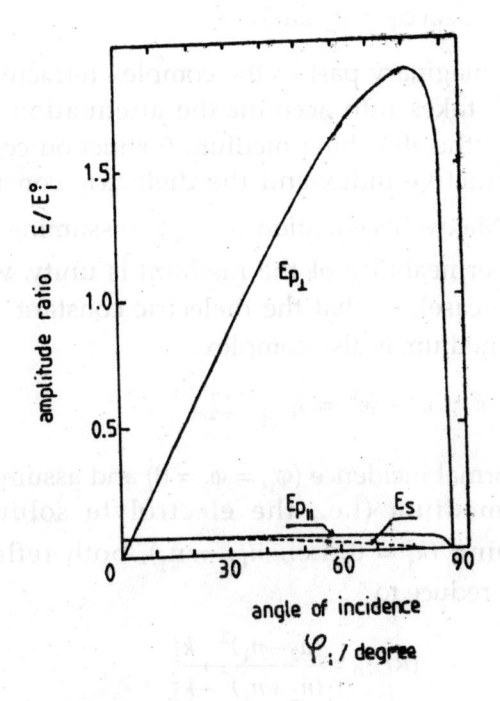

Figure 11. Dependence of E_p and E_s on angle of incidence j_i, for highly reflecting metals, in the near infrared.

In the absence of any film at the electrode/electrolyte interface (two-phase model), the reflectivity coefficients R_s and R_p, for s- and p-polarization, respectively, are expressed as follows:

$$R_s = \frac{\sin^2(\widetilde{\varphi}_1 - \widetilde{\varphi}_2)}{\sin^2(\widetilde{\varphi}_1 + \widetilde{\varphi}_2)} \tag{30a}$$

$$R_p = \frac{\tan^2(\widetilde{\varphi}_1 - \widetilde{\varphi}_2)}{\tan^2(\widetilde{\varphi}_1 + \widetilde{\varphi}_2)} \tag{30b}$$

where the angle of incidence $\widetilde{\varphi}_1$ and the angle of refraction $\widetilde{\varphi}_2$ are complex quantities, since they are related by Descartes' law (or Snell's law):

$$\tilde{n}_1 \sin \tilde{\varphi}_1 = \tilde{n}_2 \sin \tilde{\varphi}_2 \tag{31}$$

where the imaginary part of the complex refractive index $\tilde{n} = n - ik$ takes into account the attenuation of light intensity by the absorbing medium (extinction coefficient k). The refractive index and the dielectric constant are related by Maxwell's equation $\tilde{n} = \sqrt{\tilde{\varepsilon}}$ (assuming that the magnetic permeability of the medium is unity, which is usually the case), so that the dielectric constant $\tilde{\varepsilon}$ of an absorbing medium is also complex:

$$\tilde{\varepsilon} = \varepsilon' - i\varepsilon'' = \tilde{n}^2 \begin{cases} \varepsilon' = n^2 - k^2 \\ \varepsilon'' = 2nk \end{cases} \tag{32}$$

For normal incidence ($\varphi_1 = \varphi_2 = 0$) and assuming that the first medium (i.e., the electrolyte solution) is nonabsorbing ($k_1 = 0$; i.e., $\tilde{n}_1 = n_1$), both reflectivity coefficients reduce to

$$(R)_{\varphi=0} = \frac{(n_2 - n_1)^2 + k_2^2}{(n_2 + n_1)^2 + k_2^2} \tag{33}$$

For the pseudo-Brewster angle $\tilde{\varphi}_{1B}$, defined by $\tan \tilde{\varphi}_{1B} = \tilde{n}_2 / \tilde{n}_1$, so that $\tan(\tilde{\varphi}_1 + \tilde{\varphi}_2) \to \infty$, the reflectivity coefficient R_p for p-polarization as a function of the angle of incidence φ_i, passes through a minimum.

In the presence of an adsorbed layer or of an absorbing film, of thickness d and complex refractive index $\tilde{n}_2 = n_2 - ik_2$, at the interface between a nonabsorbing electrolyte ($\tilde{n}_1 = n_i$) and an absorbing metal electrode ($\tilde{n}_3 = n_3 - ik_3$) (three-phase model), there is a relative reflectivity change $\delta R/R^0$ at the interface, normalized as follows:

$$\left(\frac{\delta R}{R}\right)_{s,p} = \left(\frac{R(d) - R(0)}{R(0)}\right)_{s,p} = \left(\frac{I(d) - I(0)}{I(0)}\right)_{s,p} \tag{34}$$

where $R(d)$ and $I(d)$ are the reflectivity coefficient and the reflected light intensity, respectively, in the presence of the film.

Assuming that the film thickness d is much smaller than the wavelength of the lights, λ (i.e., $d/\lambda \ll 1$), McIntyre and Aspnes calculated the normalized reflectivity change for both s- and p-polarization:

$$\left(\frac{\delta R}{R}\right)_s = \frac{8\pi d \sqrt{\varepsilon_1} \cos\varphi_1}{\lambda} \cdot \mathrm{Im}\left(\frac{\tilde{\varepsilon}_2 - \tilde{\varepsilon}_3}{\varepsilon_1 - \tilde{\varepsilon}_3}\right) \quad (35a)$$

$$\left(\frac{\delta R}{R}\right)_p = \frac{8\pi d \sqrt{\varepsilon_1} \cos\varphi_1}{\lambda} \cdot \mathrm{Im}\left\{\left(\frac{\tilde{\varepsilon}_2 - \tilde{\varepsilon}_3}{\varepsilon_1 - \tilde{\varepsilon}_3}\right)\right.$$

$$\left. \times \left(\frac{1-(\varepsilon_1/\tilde{\varepsilon}_2\tilde{\varepsilon}_3)(\tilde{\varepsilon}_2+\tilde{\varepsilon}_3)\sin^2\varphi_1}{1-(1/\tilde{\varepsilon}_3)(\varepsilon_1+\tilde{\varepsilon}_3)\sin^2\varphi_1}\right)\right\} \quad (35b)$$

where $\mathrm{Im}(x)$ stands for the imaginary part of the complex function x.

For normal incidence ($\varphi_{i'} = 0$), both expressions reduce to

$$\left(\frac{\delta R}{R}\right)_{\varphi_1=0} = 8\pi\sqrt{\varepsilon_1}\frac{d}{\lambda}\left(\frac{(\varepsilon_1-\varepsilon_2')\varepsilon_3''(\varepsilon_1-\varepsilon_3')\varepsilon_2''}{(\varepsilon_1-\varepsilon_3')+\varepsilon_3''^2}\right) \quad (36)$$

This simplified expression allows the estimation of the relative reflectivity change at a wavelength l = 800 nm due to the presence of a Pb monolayer ($\varepsilon_2' = -15$, $\varepsilon_2'' = 20$, $d = 4$ Å) formed by under potential deposition on a silver electrode ($\varepsilon_3 = -24$, $\varepsilon_3'' = 1$) in contact with an aqueous electrolyte of $\varepsilon_1 = 1.77$. (Ref. 132). Equation (36) gives $(\delta R/R)_{\varphi_1=0} \approx 1\%$, which is in good agreement with experimental results.

For highly conductive metals ($k_3 > 10$; i.e., $\tilde{\varepsilon}_3 \gg \varepsilon_2 > \varepsilon_1$) and taking $\varepsilon_1 = 1$, Eqs. (35) can be further simplified:

$$\left(\frac{\delta R}{R}\right)_s \approx 0 \tag{37a}$$

$$\left(\frac{\delta R}{R}\right)_p \approx 8\frac{\pi d \sin\varphi_1 \tan\varphi_1}{\lambda} \cdot \text{Im}(-1/\tilde{\varepsilon}_2) \tag{37b}$$

This shows that only the *p*-component of the electric field vector plays an active role in light absorption by a film at the surface of a conducting medium and justifies the "surface selection rule." Therefore, *p*-polarizers are usually used in the experimental setup, to eliminate completely any *s*-polarization.

If the absorbing film is not too highly absorbing, that is, for a relatively small extinction coefficient k_2, Eq. (37b) reduces to

$$\left(\frac{\delta R}{R}\right)_p \approx \frac{6\pi k_2 d \sin\varphi_1 \tan\varphi_1}{\lambda} \approx \frac{4 \sin\varphi_1 \tan\varphi_1}{n_2^3} \cdot \alpha_2 d \tag{38}$$

where $\alpha_2 = 4\pi k_2/\lambda$ is the absorption coefficient of the film [see the Beer-Lambert law, Eq. (4)].

From Eq. (38), it follows that the relative change δA in the absorbance A due to the presence of the film is

$$\delta A \approx -\left(\frac{\delta R}{R}\right)_p - \frac{4 \sin\varphi_1 \tan\varphi_1}{n_2^3} \cdot A_2 \tag{39}$$

where $A_2 = \alpha_2 d$ is the absorbance of the film. This shows that, for high angles of incidence ($\varphi_1 > 80^0$), the reflectance experiment is easier to perform than the transmittance experiment, since the factor $4 \sin\varphi_1 \tan\varphi_1/n_2^3$ which is greater than unity, behaves as an enhancement factor.

This last equation allows the evaluation, in the infrared range, of the relative reflectivity change of a platinum electrode at $\lambda = 5$ µm due to a monolayer of a strong IR absorber, such as adsorbed CO, to about 1%, which is easily detectable.

(c) *Symmetry of the adsorbent sites and vibrational modes of adsorbates. Spectroscopic selection rules*

Since the first in situ infrared spectroscopic experiments have already been done on single-crystal electrodes [see the discussion in Sections II.4(i) and III. 1] and since they have showed "structural effects," it is of primary importance to consider the symmetry properties of the adsorbent/adsorbate systems. The aim of this section is just to give some general ideas about the problem and its implications for interfacial electrochemistry. The reader who wants to know more about the subject is invited to refer to the abundant literature related to the gas phase.

The recent improvement of infrared reflectance spectroscopy at the electrode/solution interface (higher sensitivity and shorter spectral accumulation times) makes it now theoretically possible to carry out a much more rigorous analysis of spectroscopic data by application of the selection rules based on symmetry properties. To gain a better knowledge of the way in which species are adsorbed onto the electrode surface is not only a challenge for academic research but is also of interest for practical applications in electrocatalysis, in that both the activity (of the surface) and the reactivity (of adsorbates) involve symmetry considerations. Obviously, a lot of new information should be available with the help of reflectance spectroscopy, at least in the near future.

Let us start by considering a species in the gas phase. Its N atoms correspond to $3N$ kinetic degrees of freedom. Classically, the translation of the center of mass accounts for three degrees of freedom, and rotation accounts for three as well, but only for nonlinear molecules (linear molecules have only two rotational degrees of freedom). Such species have low symmetry and, normally, a number of vibrational motions equal to the number of vibrational degrees of freedom, that is, $3N-6$ or $3N-5$ for nonlinear and linear molecules, respectively.

When the same species are adsorbed onto a metal, the three translational motions are not possible anymore, which results in an extra three vibrational degrees of freedom. In other words, the translational motions are converted into vibrations of the species at the surface. It follows that the degeneracy of the vibration modes due to adsorption leads to a higher symmetry ($3N-3$ or $3N-2$ vibrations, if rotation is still possible, or $3N$ if rotational motion is also converted into vibrations). In fact, the exact description of vibrational motions depends strongly on the adsorption mode, and it is easy to see that *physisorption* will affect very little the vibrational modes defined in the gas phase, while *chemisorption* may involve large structural changes of the adsorbates and therefore major changes in the vibrational frequencies.

Furthermore, at the interface, there is a reduction in the number of symmetry elements. Thus, improper rotation axes as well as centers of inversion have no meaning, and only the mirror planes perpendicular to the surface and the rotation axes normal to the surface would have to be considered.

For instance, starting from the atomic arrangements of an fcc metal, the only point symmetry groups that apply to the (100), (110), and (111) crystallographic planes are C_{6v}, C_{4v}, C_{3v}, C_{2v}, C_6, C_4, C_3, C_2, C_3, C, and C_1. In fact, the overall symmetries of unit meshes are determined by their geometry, that is, square for (100) rectangular for (110), and

Study of the Adsorb Species

hexagonal for 010. However, more important for the adsorbent-adsorbate interaction is the definition of symmetry elements (or point groups) related to bare sites, such s on-top, bridge, threefold hollow, and fourfold hollow sites, according to the usual nomenclature. These sites are represented in Fig. 12, and their symmetry elements are listed in Table 6.

In considering Table 6, it is worth stressing that:

(i) because of surface crystallography, point groups C_5 and $C_{n>6}$ have no significance;

(ii) the same types of sites are not equivalent along the different unit meshes. Thus, on-top sites will have different

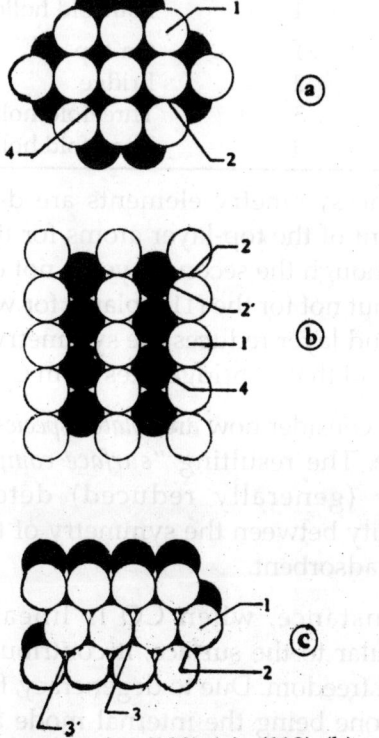

Figure 12. Site symmetries for (100) (a), (110) (b), and (111) (c) crystallographic planes. See Table 6 for identification of the sites.

symmetry properties determined by the point groups C_{4v}, C_{3v}, and C_{2v}, respectively, for the (100), (111), and (110) planes;

Table 6
Bare Sites for fcc Surfaces and Symmetry Elements

Unit mesh	Site numbering	Type of site	Point group
(100)	1	On-top	C_{4v}
	2	Bridge	C_{2v}
	4	Fourfold hollow	C_{4v}
(110)	1	On-top	C_{2v}
	2	Bridge	C_{2v}
	2	Bridge	C_{2v}
	4	Fourfold hollow	C_{2v}
(111)	1	On-top	C_{3v}
	2	Bridge	C_s
	3	Threefold hollow	C_{3v}
	3'	Threefold hollow	C_{3v}

(iii) the symmetry elements are determined by the arrangement of the top-layer atoms for the (100) and (110) planes (although the second layer is not directly below the first one), but not for the (111) plane, for which the presence of the second layer reduces the symmetry of top sites from C_{6v} to C_{3v} and that of bridge sites from C_{2v} to C_s.

Let us consider now an *isolated species adsorbed* on a *bare surface site*. The resulting "surface complex" will have a symmetry (generally reduced) determined by the compatibility between the symmetry of the adsorbate and that of the adsorbent.

For instance, when CO is linearly bonded and perpendicular to the surface, it contributes six degrees of vibrational freedom. Due to degeneracy, four vibrations are observed, one being the internal mode and three coming

Study of the Adsorb Species

from translation. Thus, the symmetry is reduced from $C_{\infty v}$ (free molecule) to C_{4v} if the CO molecule is adsorbed at on-top sites of an fcc (100) unit mesh, leading to the four vibrational modes represented in Fig. 13. In C_{4v} symmetry, v_3 and v_4 form a degenerate pair of vibrations. However, there is a lifting of the degeneracy if the adsorption occurs at on-top sites of (111) or (110) fcc surfaces, since the symmetry groups become, respectively, C_{3v} and C_{2v} (cf. Table 6). Similarly, adsorption onto bridge sites would lead to a complex with C_{2v} symmetry on (100) and (110) planes (Fig. 13), but reduced to C_s on the (111) plane. If the CO molecule is tilted, an asymmetric bridging would result, as a consequence of which the point group would become C_s on all planes.

Apart from CO, two important adsorbed species in electrocatalysis (see the results given in Section III.1) are formate and methoxy species. While the former is planar and has C_{2v} symmetry, the latter has a C_{3v} symmetry determined by the C–H bonds–with all the consequences of this symmetry with respect to adsorption on fcc metals. Figure 14 illustrates this particular situation in the case of the complex formed by a methoxy species on a (100)

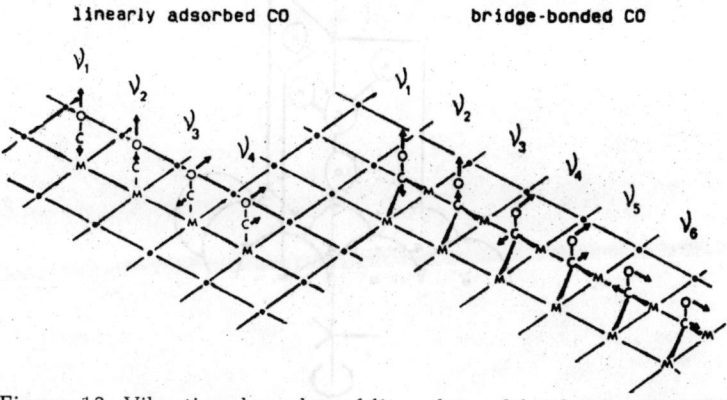

Figure 13. Vibrational modes of linearly and bridge-bonded CO adsorbed on an fcc (100) plane.

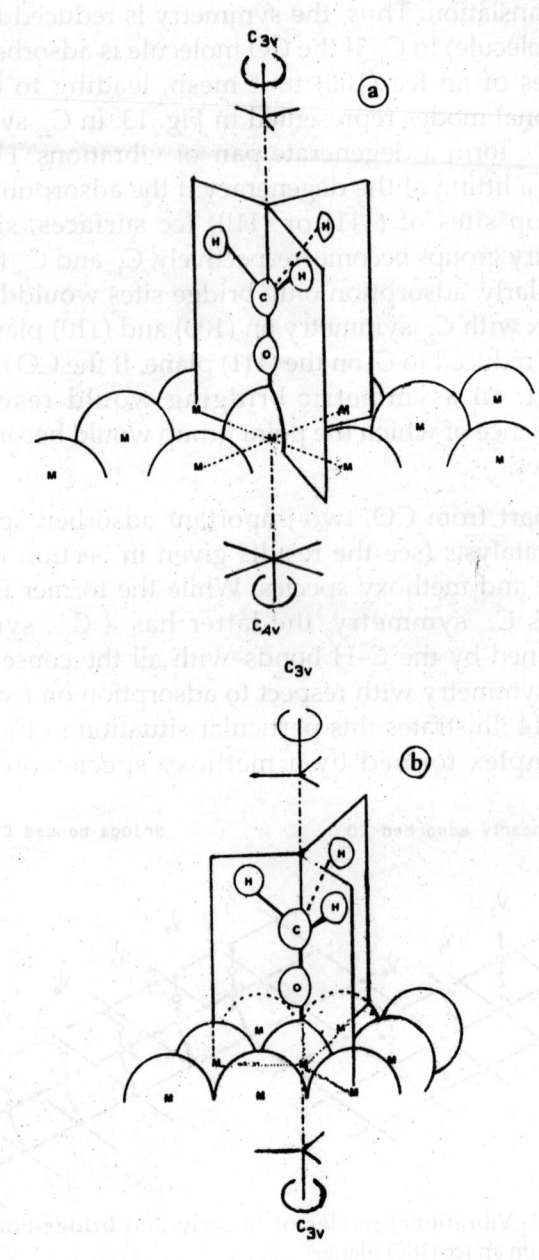

Figure 14. Methoxy adsorption at (a) C_{4v} and (b) C_{3v} sites.

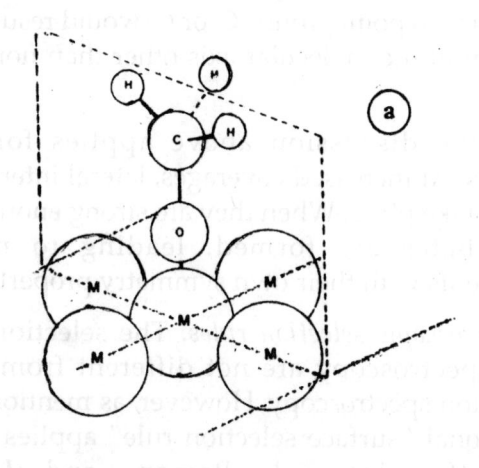

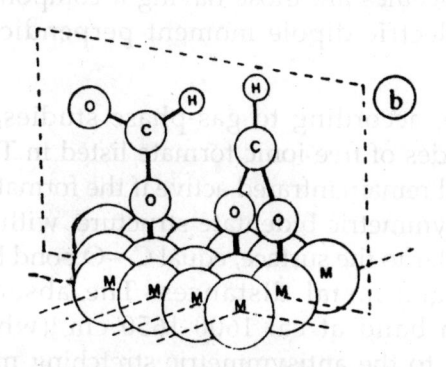

Figure 15. Adsorbed species at C_{4v} sites: (a) methoxy group; (b) monodentate and bidentate formate.

surface. It is clear that the symmetry of the methoxy species is reduced to C_s, due to incompatibility between the point groups C_{4v} of on-top sites and C_{3v} of the adsorbate, while the symmetry of the complex formed between formate and the surface is C_{2v} when the formate group is bonded to the surface via the two oxygens, and only C_s if it is singly bonded to the surface (Fig. 15). Of course, further reduction in symmetry to point groups C_s or C_1 would result from any orientation of the molecular axis other than normal to the surface.

All the discussion above applies for isolated adsorbates. At increased coverages, lateral interactions are known to take place. When they are strong enough, islands of adsorbates are formed, leading to new local arrangements with their own symmetry properties.

Spectroscopic selection rules. The selection rules for surface spectroscopy are not different from those for transmission spectroscopy. However, as mentioned above, an additional "surface selection rule" applies to surface species (for infrared, Raman, and UV-visible spectroscopies). The rule states that the only observable surface molecules are those having a component of their dynamic electric dipole moment perpendicular to the surface.

Hence, according to gas-phase studies, of the six normal modes of free ionic formate listed in Table 7, only three would remain infrared active if the formate adsorbate had a C_{2v} symmetric bidentate structure, with the C_2 axis perpendicular to the surface, equal C—O bond lengths, and equal oxygen-metal distances. The absence of any absorption band at ca. 1600-1650 cm^{-1} (which would correspond to the antisymmetric stretching mode of free formate ions) may be taken as proof that the formate adsorbate is oriented with the two oxygen atoms toward the surface, as indicated above. However, recent electrochemically modulated infrared reflectance spectroscopy (EMIRS) investigations at an electrolyte solution/smooth platinum electrode interface led us to conclude that the rather complex band which extends from 1620 to 1700 cm^{-1}, and which includes the δ(HOH) deformation mode of water at 1625 cm^{-1}, may also contain

Table 7
Vibrational Modes of Formate

Mode		Description	Vibrational frequency (cm^{-1}) in aqueous solution[a]	Infared activity in the C$_{2v}$ configuration
H | C / \\ O→ ←O	OCO bend	d(OCO)	772	Yes
H + | C / \\ O O	CH out-of-plane bend	p(CH)	1073	No
H | C / \\ O O	OCO symmetric stretch	V_s(COO)	1366	Yes
H→ | C / \\ O→ ←O	CH in-plane bend	d(CH)	1377	No
H | C / \\ O→ ←O	OCO anatisymmetric stretch	v_a(COO)	1567	No
H↑ C / \\ O→ ←O	CH stretch	v(CH)	2841	Yes

near 1640 cm^{-1} a contribution from the antisymmetric stretch of formate [see also Section III.1(iii)]. The likely surface configuration that would result in such a mode becoming infrared active would be a formate species oriented with only one oxygen toward the surface.

Similarly, in the case of methanol adsorption, the multiplicity of C-H stretches, as observed by EMIRS in the range 2800-3100 cm^{-1}, could be interpreted as due to different orientations at the surface. Only one mode, due to the C—H symmetric stretch, should be active if the C—O axis is perpendicular to the surface, whereas any tilt from this position would make the asymmetric stretching modes also active.

Thus, the potential utility of infrared reflectance spectroscopy, even though the technique is still difficult, for understanding the mechanisms of electrocatalytic reactions has been demonstrated. It should help, for instance, in understanding why the reactivity of methanol (or methoxy species) is so different from that of formic acid on various catalytic metals and on different crystallographic planes of the same metal.

(d) Surface spectroscopy in the UV-visible range

The first attempts to use reflectance spectroscopy in the UV-visible range were done, with the aim of understanding the structure of the electrode/electrolyte interface. However, only limited information readily usable for electrocatalysis was obtained.

Real interest in this approach came a few years later when it was demonstrated that optical studies (i) were sensitive enough to characterize single-crystal electrodes, (ii) could help in understanding the formation of passive films on non-noble metals, (iii) could be used to investigate the formation of adsorbed hydrogen atoms on smooth pt,12' and, more surprisingly, could be used to study CO adsorption on catalytic metals by following the changes in the formation of their oxide layers.

Several extensive reviews are available in the literature, to which the reader may refer for experimental details. Since the beginning of reflectance spectroscopy, various experimental setups have been developed, either using modulation techniques with phase-sensitive detection or signal processing and averaging with the help of computers.

Briefly, potential-modulated reflectance spectroscopy (PMRS), the most commonly used technique, is an in situ technique in which periodic reflectivity changes (δR) in the monochromatic light reflected from the electrode surface (whose reflectivity is R) are produced by a sine or square-wave potential modulation and detected by a lock-in amplifier. By this technique, typical $\delta R/R$ values of 10^{-5} can be routinely obtained, which is generally sufficient for most studies.

However, the PMR technique itself has its own limitations. If slow processes occur on the electrode surface, the frequency of the electrochemical response may fall below the hertz range, which is usually the lower limit for normal operation of phase-sensitive amplifiers. The situation is even worse with non-noble metals, at the surface of which totally irreversible processes are known to occur, or with single crystals, which may reconstruct under potential modulation. Moreover, the interpretation of PMR spectra is often difficult. Recently, it was suggested that the maxima might correspond to the normal absorption bands of species, which, if confirmed, would make PMRS interesting as an analytical tool.

An alternative technique is to use rapid-scan spectrometers coupled with high-speed signal averaging.

The very short time necessary for acquisition of a single spectrum allows the use of on-line data-recording systems to collect and average a large number of spectra in a few seconds or a few tens of seconds. Thus, signal-to-noise ratio improvement is obtained, and sensitivities of 10^{-3} (in $\delta R/R$) are currently achieved. In differential reflectance spectroscopy (DRS), series of a few hundred spectra are usually collected at different potentials. The subtraction of the averaged series, one serving as reference, gives differential spectra which allow the dependence of the layer on potential to be followed. Alternatively, the potential may be kept constant during the successive series. Thus, after subtraction of the reference series, the time dependence of the phenomenon is observed. Such information is particularly useful for-studying the growth of metal oxide layers or corrosion processes.

The new generation of rapid-scan spectrometers with diode arrays combined with a polychromator and multichannel analyzer certainly offers interesting possibilities for DRS. Some applications have been described.

A very different approach to optical measurements for the study of electrocatalysis was developed by Beden et al., on the basis of the coupling of cyclic voltammetry with UV-visible reflectance spectroscopy (CVUVRS). The technique is based upon the recording of "reflectograms" synchronously to the voltammograms. The reflectograms represent the changes of reflectivity $\delta R/R$ versus the electrode potential E, at fixed wavelengths. Once a series of reflectograms is recorded, the reflectivity changes are plotted in a three-dimensional diagram $\delta R/R = f(E, \lambda)$ which, if cut at constant E, yields the usual reflection-absorption spectra $\delta R/R = f(\lambda)_E$. Despite the duration of the experiments, the above technique is particularly suitable for interfacial electrochemical investigations, in that cyclic voltammetry provides a permanent control of the surface state during spectral acquisition, which the other techniques cannot offer. CVUVRS has been successfully applied to the study of absorbed organic species on catalytic electrode surfaces as well as to follow the changes that may affect unstable surfaces, such as oxide layer growth or corrosion processes. CVUVRS has also been very successfully employed to investigate the state and valency of oxide layers at noble metal electrodes. This is a very important point in electrocatalysis, since the extra oxygen atoms needed to completely oxidize the absorbed oxygenated organic molecules to CO_2 can come, in aqueous solution, from adsorbed water and hydroxyl species or from oxygen and oxide layers. As a typical example, the nature of the oxide layers which are involved in the electrocatalytic oxidation of formate at rhodium (and which contribute to the oscillating behavior encountered) was elucidated by CVUVRS.

(e) Surface spectroscopy in the infrared range

The case of external reflectance infrared spectroscopy is more complicated, because of the strong absorption of infrared radiation by aqueous solutions, so that the technique was long considered by spectroscopists to be infeasible.

The first in *situ* experiments, reported workers in 1980 and 1981, demonstrated the feasibility of a now known as electrochemically modulated infrared technique now known as electrochemically modulated infrared reflectance spectroscopy (EMIRS) (Fig. 16a). Since this initial work, many improvements have been made, either by modifications of the technique or by enlarging considerably the variety of subjects investigated. Several extended reviews and chapters in books may be consulted by the reader interested in the technical details.

The variants of infrared reflectance spectroscopy differ type of spectrometer employed (dispersive or Fourier transform), the reflection mode (external or internal), and, principally, the technique used for signal enhancement, without which no signals due to adsorbed species can be extracted. In Table 8 the main infrared reflectance techniques are listed.

In all infrared reflectance techniques, data processing plays an important role. The tasks of data processing are:

● to sort out the information that corresponds to surface species from that arising from bulk species from that arising from bulk species (generally much more intense); and

● to improve the weak signal-to-noise ratio (S/N) inherent to this type experiment.

Techniques coupled with dispersive instruments. Except in some pioneering experiments, the most powerful technique, EMIRS, consists in modulating the electrode potential between two limits (the lower potential, E_c, at which the reflectivity is R_c, and the upper potential, E_a, at which the reflectivity is R_a), with a modulation amplitude of 50 to 500 mV and a frequency of a few hertz. By synchronous analysis of the signals, it is possible to reject the nonmodulated information (due to absorption by species in the electrolytic solution) and to amplify the signal-to-noise ratio of the absorption bands due to vibrations of adsorbed species. In fact, a frequency of about 8 to 15 Hz is a good compromise that allows the electrochemical system to respond to the change in potential and the synchronous detection to work satisfactorily. The dc output of the phase-sensitive detector, $\delta R(\bar{v})$, at a wavenumber $\bar{v}$ is then stored, averaged if necessary, and divided by the electrode reflectivity $R(\bar{v})$ in order to obtain the spectrum

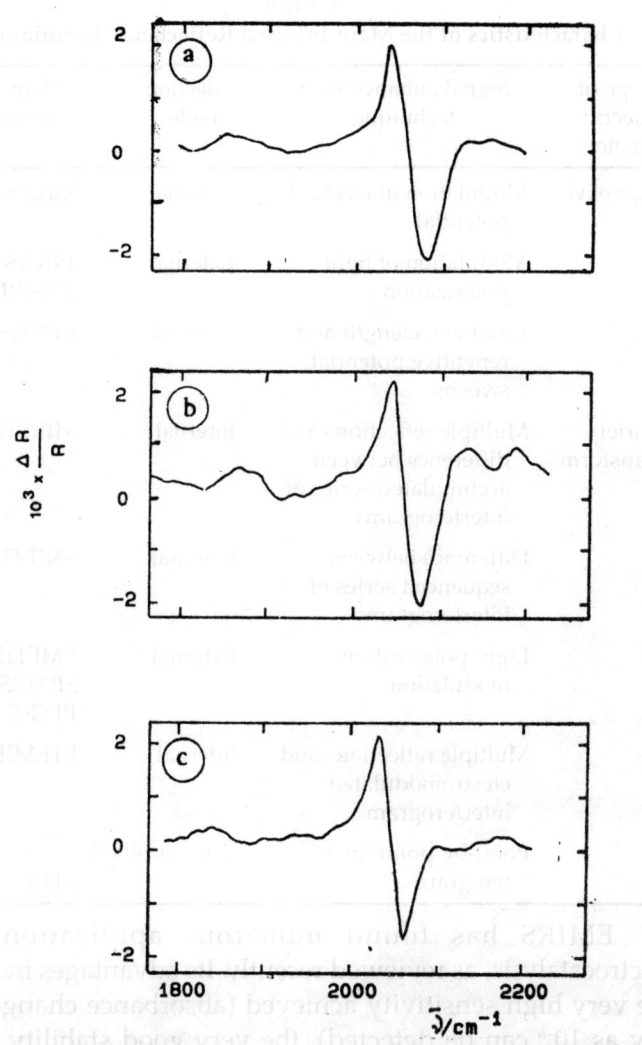

Figure 16. EMIRS spectra of CO species resulting from the adsorption on a Pt electrode (in 0.5 M HClO$_4$ at room temperature) of: (a) 0.25 M CH$_3$OH; (b) gaseous CO; (c) 0.25 M HCOOH.

in its final dimensionless form ($\delta R/R$, $\bar{v}$). The technique, which was introduced by Bewick et al., is fully described.

Table 8
Characteristics of the Main Infrared Reflectance Techniques

Type of spectrometer	Signal enhancement technique	Reflection mode	Name of technique
Dispersive	Modulation of electrode potential	External	EMIRS
	Modulation of light polarization	External	IRRAS PM-IRRAS
	Fixed wavelength and repetitive potential sweeps	External	LPSIRS
Fourier transform	Multiple reflections and difference between accumulated series of interferograms	Internal	MIRFTIRS
	Difference between sequenced series of interferograms	External	SNIFTIRS
	Light polarization modulation	External	PMFTIRRAS SPAIRS PDIRS
	Multiple reflections and electromodulated interferogram	Internal	FTEMIRS
	Point-by-point interferogram	Time-resolved	FTIR

EMIRS has found numerous applications in electrocatalysis, as reviewed recently. Its advantages include the very high sensitivity achieved (absorbance changes as low as 10^{-6} can be detected), the very good stability with time even without purging gas, and the very good stability with time even without p comparatively low cost of the equipment relative to Fourier transform spectrometers. Figure 17a illustrates the case of methanol chemisorption on platinum electrodes in acid medium. Different EMIRS bands are seen. They correspond to various surface species, including formate and carbon monoxide. Furthermore, with the same basic modular equipment, it is possible to set up other types of techniques such as linear potential sweep infrared spectroscopy (LPSIRS), introduced by Kunimatsu, which allows the investigation, at fixed wavelengths, of the species produced at different electrode potentials during the voltammetric sweep. On the other and, there are some inconveniences associated with EMIRS. The most important one is that the electrode potential is not fixed at a given value (because of potential modulation, whose amplitude can reach 0.5 V), so that many electrochemical processes may occur in the potential range of modulation, leading to highly complex EMIRS spectra. The second one is that the modulation technique combined with the so-called Stark effect (shift of frequency with potential) leads in most cases to signals having a pseudo-derivative shape, which makes it difficult to extract quanti-tative information. The third one is that the commercial grating spectrometers specially designed for EMIRS, for many technical reasons (related to infrared materials, wavelength scanning rates, change of filters for sorting out the grating orders, computerized data acquisition and processing), do not allow the whole range of 400-4000 cm^{-1} to be covered in a single experiment.

Alternatively, a very informative technique is to modulate the polarization state of the infrared radiation, instead of the potential. Derived from IRRAS, originally developed for investigations at the solid/gas interface, the technique is a direct application of theory [see Section II.3(iv.b), on s- and p-polarizations] and was adapted to investigations at the electrode surface, to confirm EMIRS results by quantitative measurements and to obtain conformational information with respect to the orientation of adsorbates. Recent applications have shown this technique to have very good sensitivity.

Techniques coupled with Fourier transform spectrometers. By alternatively coadding short series of interferograms at two given electrode potentials, it is possible to take advantage of the very high luminosity of Fourier transform infrared spectrometers and to obtain, after subtraction and ratioing, spectra of adsorbed species of the form

$$\frac{\delta R}{R} = \frac{R_a - R_c}{R_c} = \frac{R_a}{R_c} - 1$$

Coupled with an external reflection spectroelectrochemical cell, the technique is called subtractively normalized interfacial Fourier transform infrared spectroscopy (SNIFTIRS) and was developed by Pons *et al.* The successive storage of interferograms at E_c and E_a allows drifts, whatever their electrical or chemical origin, to be minimized. Thus, long-duration experiments are possible, up to several hours. Typically, 50 interferograms are stored in a few minutes at each potential limit. Once the potential is switched, data accumulation is only triggered after the short latency time necessary for the capacitive currents to go to zero. Figure 17b gives an example of a SNIFTIRS spectrum. The strongest peaks correspond to solution species (perchlorate ions, methanol, CO_2, and probably formic acid). Surface species (CO and formate) give rise to comparable intensities as in the EMIRS experiment of Fig. 17a.

A variant uses a coupling of Fourier transform spectroscopy with light polarization modulation (PMFTIRRAS). The frequency of the photoelastic modulator (70 kHz) is much higher than that of the trigger signal used in recording the interferograms, so that, using phase-sensitive detection, it is possible to demodulate the detected signal before its Fourier transformation. Actually, the resulting spectra are differences between spectra recorded with sand p-polarized infrared light and are supposed to

Study of the Adsorb Species

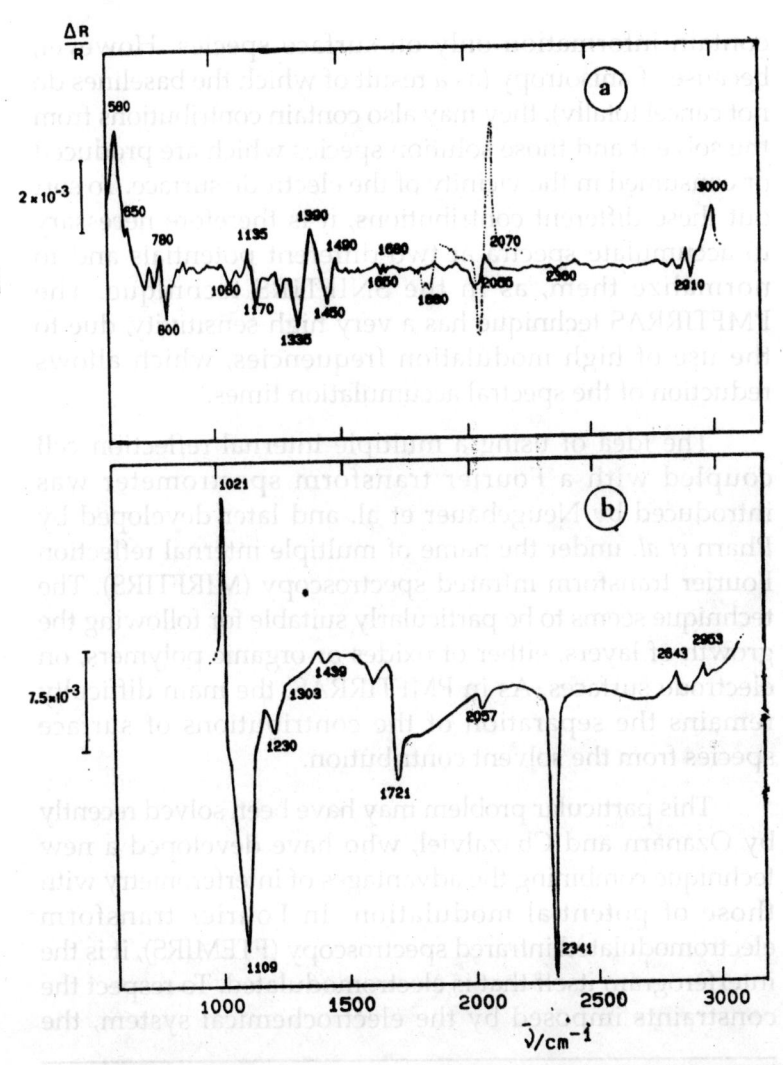

Figure 17. Infrared reflectance spectra of the layer formed by chemisorption of CH_3OH from perchloric acid medium (room temperature) onto a polycrystalline platinum electrode. (a) EMIRS spectrum of the only surface species resulting from chemisorption of 10^{-3} M CH_3OH. CO (band at ca. 2050 cm^{-1}) and formate (bands at ca. 1450 cm^{-1}) result from oxidation. The strong band at ca. 580 cm^{-1} corresponds to Pt—O or Pt-OH stretches. Increasing the concentration

contain information only on surface species. However, because of anisotropy (as a result of which the baselines do not cancel totally), they may also contain contributions from the solvent and those solution species which are produced or consumed in the vicinity of the electrode surface. To sort out these different contributions, it is therefore necessary to accumulate spectra at two different potentials and to normalize them, as in the SNIFTIRS technique. The PMFTIRRAS technique has a very high sensitivity, due to the use of high modulation frequencies, which allows reduction of the spectral accumulation times.

The idea of using a multiple internal reflection cell coupled with a Fourier transform spectrometer was introduced by Neugebauer et al. and later developed by Pharn *et al.* under the name of multiple internal reflection Fourier transform infrared spectroscopy (MIRFTIRS). The technique seems to be particularly suitable for following the growth of layers, either of oxides or organic polymers, on electrode surfaces. As in PMFTIRRAS, the main difficulty remains the separation of the contributions of surface species from the solvent contribution.

This particular problem may have been solved recently by Ozanarn and Chazalviel, who have developed a new technique combining the advantages of interferometry with those of potential modulation. In Fourier transform electromodulated infrared spectroscopy (FTEMIRS), it is the interferogram itself that is electromodulated. To respect the constraints imposed by the electrochemical system, the

---

of CH_3OH in solution leads to an increase in surface CO (dotted band). (b) SNIFTIRS spectrum of both absorbed and solution species resulting from oxidation of 0.1 M CH_3OH. By comparison with panel a, it is clear that strong solution bands, such as those corresponding to prechlorate (at ca. 1100 cm^{-1}) or possibly to formic acid (C=O stretch at ca. 1720 cm^{-1}), are detected as well. However, the band intensities for surface species have approximately the same intensities as in the EMIRS experiment of panel a.

mobile mirror of the interferometer had been designed to move at the extremely slow speed of 6 µm s^{-1}. Hence, the electrode potential can be modulated at a frequency of around 100 Hz, compatible with the response of the electrochemical reaction. At the Si-aqueous electrolyte interface, using an ATR electrochemical cell with 15 reflections, signals as weak as 10^{-6} (in absorbance units) could be detected over the entire range 800-4200 cm^{-1}, with 10-cm^{-1} resolution and after only one hour of spectral accumulation. No application to electrocatalysis has been published yet.

Also of great interest, for both fundamental research and applied studies, is the time dependence of electrochemical processes. Transient species often play a kinetic role as reaction intermediates and may be involved in the rate-determining step. However, the first infrared reflectance experiments, because of too long spectral accumulation times, were far beyond the time scale required for identification of transient intermediates.

Recently, attempts have been made to use single potential alteration infrared spectroscopy (SPAIRS) with reduced spectral acquisition times of a few seconds. These "real-time FTIR experiments are a good approach to the time dependence problem and allowed the authors to follow some slow adsorption processes on electrode surfaces. The first experiments involving time-resolved spectroscopy on a millisecond scale were reported by Pons and co-workers. In their experiments, -the interferogram is collected as a set of discrete points, the moving mirror being displaced step by step. A final interferogram is reconstructed from the averaged intensities at each point and then Fourier transformed. Interesting applications of the technique have been made in the far infrared, where the detection of metal-metal vibration was reported.

(v) Other Spectroscopic and Diffraction Techniques

In the past ten years, most investigations of the electrode/electrolyte interface have been done by infrared reflectance spectroscopy, using the versatile EMIRS or SNIFTIRS techniques [see Section II.3(iv.e)]. The impact of these techniques has been considerable, in terms of the importance of the results, which in many cases led to new concepts, and the variety of electrochemical systems investigated so far, ranging from electrocatalysis to electrosynthesis or corrosion.

However, insofar as applications to interfacial electrochemistry and electrocatalysis are concerned, many other techniques, involving absorption, reflection, or diffraction of electromagnetic waves or of electrons, are interesting as well. These will now be reviewed briefly.

(a) Raman spectroscopy

Raman spectroscopy was recognized as a useful tool for surface investigation after the original discovery in 1974 by Fleischmann and co-workers that the Raman spectrum of pyridine adsorbed on a silver electrode in aqueous KCl was considerably amplified if the electrode was previously subjected to oxidation-reduction cycles (ORC) under potentiostatic control. Since then, surface enhanced Raman scattering (SERS) has been observed with different active metals, and numerous papers have been devoted to the enhancement mechanism itself and to the study of adsorbed molecules, or layers, on electrode surfaces.

While the Raman effect was originally predicted by Smekal, the technique was developed by Raman, who recorded the first spectra. Briefly, when an electromagnetic wave irradiates a molecule, the associated electric field induces a small dipole moment through its polarizability. If the source emits an intense radiation, such as that emitted by a Laser, in the visible part of the spectrum, the collision between the incident photon and the molecule is nearly elastic, and the oscillating dipole radiates light at the same frequency, but not necessarily in the same direction, leading to intense Rayleigh scattering. Additional scattering comes from the contribution of the normal modes of vibration of the molecule, which can add to or subtract from the main scattered radiation. Thus, very weak spectral bands can be detected at frequencies slightly shifted from the Rayleigh frequency, and these shifts reflect the characteristic vibrational motions of the molecule. This is the basis of normal Raman spectroscopy (NRS). However, if the frequency of the exciting monochromatic radiation is chosen so that it corresponds to an electronic absorption, amplified Raman spectra are obtained. The technique, which is called resonance Raman spectroscopy (RRS), may be useful when the NR spectra are too weak to be accurately measured. As mentioned above, a more interesting technique for electrocatalysis is SERS. Although not yet totally understood, it is now believed that the SERS effect, which can lead in some cases to enhancement factors as high as 10⁶ or more, originates when several conditions are fulfilled, namely:

(i) the molecules have to be either adsorbed (preferably chemisorbed) or in the vicinity of the electrode surface;

(ii) the metal surface has to be highly reflecting at the wavelength of the exciting radiation. In the visible, Ag, Au, and Cu, respectively, are the strongest SERS active metals, but recently the first reports have appeared of SERS with other metals, such as Ni, Pd, Al or even Hg;

(iii) the surface has to be rough, the minimum microscopic roughness being one order of magnitude greater than the atomic scale. Such a surface can be obtained by different means, the simplest process probably being electrochemical cycling. It has been shown that, in some cases, a single oxidation-reduction cycle of the surface was enough to produce an intense Raman spectrum.

Under these conditions, most adsorbed molecules can give rise to SERS, but with quite different amplification effects, as pointed out by Otto and co-workers.

Experimentally, the basic equipment for SERS is the same as for NRS or RRS. Laser sources are used to provide enough power for weak Raman signals to be detected. High discrimination against stray light (due to the intense Rayleigh band) is necessary, which can be achieved using double or triple monochromators. Multichannel detection is now often preferred to the photomultiplier tubes employed in the original experiments. A recent and detailed discussion of experimental problems may be found together with a description of spectroelectrochemical cells.

Applications of SERS to different domains of interfacial electrochemistry may be found in the literature (see the references cited above). The lack of understanding of the exact nature of the amplification effect, as well as the restriction of SERS to a few active metals, has certainly limited the applicability of the technique. Thus, much effort has been directed toward understanding the enhancement effect itself, rather than toward the use of SERS as an analytical tool. However, in the past ten years, extremely valuable qualitative information has been obtained by several groups on certain systems of interest for electrocatalysis, especially those systems for which information is also available from infrared reflectance spectroscopy. Some criteria that may help in distinguishing the signals of the surface from those of the bulk have been investigated. Thus, in several articles, Weaver and co-workers have compared the band frequencies, bandwidths, and selection rules for different adsorbates on gold and silver electrodes, two metals which are SERS active and catalytically interesting. Different applications of SERS to electrochemistry may be found in Refs. 187 and 197. New developments have been reported in the field of corrosion and in the field of polymer films.

It is now true that the SERS and the infrared spectroscopic approaches are complementary for probing molecular structure of adsorbates at the electrode/ solution interface. For instance, infrared spectroscopy, which does not require surface roughness, can be applied to any type of surface, including well-defined single-crystal faces of any metal. Conversely, despite recent developments, SERS is still limited to a relatively low number of active metals. However, it has great sensitivity and is particularly suitable for investigations in the highly interesting spectral range of 100-600 cm^{-1} (where most substrate-adsorbate bonds absorb IR radiation); in this spectral range, infrared spectroscopy suffers from severe limitations due to weakness of sources, low detectivity of detectors, and increasing absorbance of materials.

Why then is Raman spectroscopy not more popular? As pointed out recently by Hendra and Mould, the reason is the common occurrence of stray absorbance and fluorescence interference. These authors estimated that, because of this problem, less than 20% of the samples that they have analyzed so far have provided usable spectra.

A new development of the technique might solve this problem. Recent papers have described Fourier transform Raman spectroscopy (FT-Raman) as a promising alternative to conventional Raman spectroscopy, especially if it is applied in the near infrared. According to several authors, not only is the new technique more versatile, but it would also be applicable to samples which normally exhibit fluorescence in conventional Raman work.

(b) Ellipsometry

Ellipsometry is an analytical technique based on the measurement of changes in the polarization state of light caused by the interaction of the incident beam with a physical system, for example, after reflection at the interface between two different media. It has been mainly employed within the UV-visible and IR spectral regions, allowing thus the in situ investigation of the electrode/ electrolyte interface.

The use of ellipsometry in interfacial electrochemistry is now relatively widespread, and several review papers have been published, to which the reader may refer in order to find discussions on the basic principles, on the instrumentation, and on the information obtained.

In classical ellipsometry, the wavelength of the incident light is fixed, and the ellipsometric parameters (ψ, Δ) are measured. Because ψ and Δ are directly related to, respectively, the ratio of the reflectivity coefficients and the difference in phase angles of p- and s-polarizations, ellipsometry is highly surface selective. Moreover, its high sensitivity makes it a powerful technique for the investigation of very thin films and subinonolayers at the electrode surface.

Depending on the spectral range used (UV-visible or IR), information on the electronic properties, particularly the values of the real and imaginary components of the complex dielectric constant, or on the vibrational states of the film is obtained. The thickness of the film can also be estimated from ellipsometry measurements. When measurements are taken over a wide range of wavelengths, in so-called "spectroscopic ellipsometry," information can be obtained on the film structure, because the ellipsometry response is equivalent to that obtained in reflection-absorption spectroscopy, with the advantage of a higher specificity for monitoring the surface film (which thus eliminates the need to obtain difference spectra by subtracting a reference spectrum that includes the spectral response of the species in solution).

Ellipsometry in the UV-visible range, with potential control for application to interfacial electrochemistry, was first developed by Reddy and Bockris in 1964. Later on, the technique was extensively used for in situ measurements of the optical constants of passive oxide layers, mainly for purposes of studying corrosion and characterizing electrochemically grown polymeric films, such as polyaniline.

The study of chemisorbed species is theoretically possible, but so far, due to the complexity of the experiments and to the difficulty of nterpretation, only investigation of H and O adsorption and of anionic adsorption in the double layer has been considered. However, ellipsometry has been much more successful in the study of upd of some metallic adatoms on metallic substrates, for example, Pb/Ag and Pb/Cu. Some attempts were also made to develop ellipsometry in the infrared range. Applications to the metal/solution interface were reported by Dignam and Baker and also by Graf et al.

(c) Nonlinear optical spectroscopy

Nonlinear optical techniques, such as second-harmoni c generation (SHG) and sum frequency generation (SFG), are highly specific techniques for studying interfaces between two centrosymmetric media. Due to its high surface sensitivity, as the result of a filtering system that rejects the bulk signal (at fundamental frequency ω), but passes the second harmonic (at frequency 2ω), SHG appears to be a powerful spectroscopic technique for investigating in situ the structure of the electrode/ electrolyte interface, and particularly submonolayers of adsorbed molecules. The first SHG studies have mainly concerned electronic transitions (due to the use of a visible laser and photomultipliers as detectors), so that they only gave information on the electronic structure of the electrode/electrolyte interface. However, the low sensitivity of infrared detectors has been overcome in the sum frequency generation technique, which uses a tunable infrared laser to probe the vibrational spectrum of the adsorbates and a visible laser to convert this spectrum to a visible spectrum by sum frequency generation. SFG can provide in situ vibrational spectra of adsorbed molecules at the electrode/electrolyte interface. This was effectively demonstrated recently in the case of the adsorption of carbon monoxide 120 at platinum in contact with 0.1 M HClO4 aqueous electrolyte.

(d) X rays

Compared to spectroscopies in the UV-visible or in the infrared range, which give information on the electronic structure of the electrode surface as well as on the molecular structure and local environment of adsorbates, in *situ* X-ray techniques have the potential ability to provide details about longer range ordering. Indeed, with the advent of appropriate sources, many different techniques have been developed in the past ten years that use either scattering or absorption of X rays, in both the transmission and the reflection mode.

As X rays are scattered by all the matter in their path, they are inherently not surface sensitive. Therefore, for surface studies, it is necessary to design special experiments with improved surface selectivity. Various approaches can be used. Table 9 gives the names of the techniques, together with the type of X-ray source, the way in which surface sensitivity is achieved, and the applicability to electrochemistry.

Study of the Adsorb Species

Table 9
In Situ X-Ray Techniques and Their Applicability to Interfacial Electrochemistry

Technique	X-ray source	Method of achieving surface sensitivity	Applicability to *in situ* investigations of Electrochemical systems.
Scattering methods			
TRBD (total reflection Bragg diffraction)	Synchrotron	Glancing angle of incidence	Seems possible but not yet tried
X-ray standing waves	Synchrotron		Only ex situ experiments were carried out
In situ X-ray diffraction	Conventional laboratory X-ray source	Position-sensitive X-ray detection with electrode potential modulation	Yes, tested on several electrochemical systems
Absorption methods			
EXAFS (extended X-ray absorption fine structure)	Synchrotron	Conventional EXAPS not surface sensitive	
Reff EXAFS	Synchrotron	Reflection below the critical angle	Seems difficult
SEXAFS (surface extended X-ray absorption fine structure)	Synchrotron	Measurements of electron yield	None
XANES (X-ray absorption near edge structure)	Synchrotron	Measurements of fluorescence yield coupled with glancing	Yes, several attempts have been made

Investigations at the electrode/solution interface are theoretically possible, since X rays at suitable wavelengths are able to penetrate a thin layer of solution. However, cell design remains difficult [see Section II.3(iv.a)]. One of the most successful techniques, initially developed by Fleischmann et al, utilizes a position-sensitive X-ray detector in conjunction with potential modulation, in a very similar way as in the modulation spectroscopies described previously. Its versatility, compared to other X-ray techniques, comes from the use of a conventional laboratory X-ray source instead of a synchrotron. Up to now, however, only a few studies with this technique have been related to electrocatalysis, the most relevant ones concerning hydrogen or carbon monoxide adsorption on platinized platinum in acid medium. It was shown that the electrode surface reconstructs upon adsorption of the so-called weakly adsorbed hydrogen or carbon monoxide (independently of its coordination state), but not with strongly adsorbed hydrogen.

The feasibility of the application of most of the other techniques listed in Table 9 to electrochemical systems has been demonstrated, generally taking corrosion of metals or underpotential deposition of metals as examples. In that sense, further progress can be expected in the field of electrocatalysis. However, the necessity for routine access to a synchrotron-type source will certainly be a serious limitation.

(e) Mass spectroscopy

The basic principle of mass spectroscopy has been described in Section II.2(ii). Electrochemical mass spectroscopy (EMS) was first developed by Bruckenstein and Comeau and later used to investigate the behavior of adsorbed CO at a Pt electrode in acid medium. A porous membrane (of small pore size and nonwetting

Study of the Adsorb Species

characteristics) is set between the porous working electrode and a fritted disk which protects the inlet of the mass spectrometer. Thus volatile products can pass freely through to the inlet to be analyzed

Later on, the system was improved by Wolter and Heitbaum, especially in terms of a considerably reduced time response, leading to the so-called DEMS technique (see Table 10). Examples of DEMS results are given in Fig. 18. It is striking how the production of CO_2 and methyl formate follows the rate of oxidation of methanol. However, the amount of the latter product is only 1% of that of the former.

Wilhelm et al. introduced ECTDMS, which combines electrochemical techniques with thermal desorption and mass spectrometry. In this case, the electrode has to be transferred from the electrochemical environment to the ultra-high-vacuum chamber, where it is heated in order for the desorption of adsorbed fragments to occur, which is followed by immediate mass analysis. Similarly, in the SIMS

Table 10
Mass Spectroscopic Techniques

Technique	Necessitates transfer to ultra-high-vacuum (UHV)	Applicable to interfacial electrochemistry
EMS (electrochemical mass spectroscopy)	No	Yes
DEMS (differential electrochemical mass spectroscopy)	No	Yes
ECTDMS (electrochemical thermal desorption mass spectroscopy)	Yes	Yes
SIMS (secondary ion mass spectroscopy)	Yes	Yes

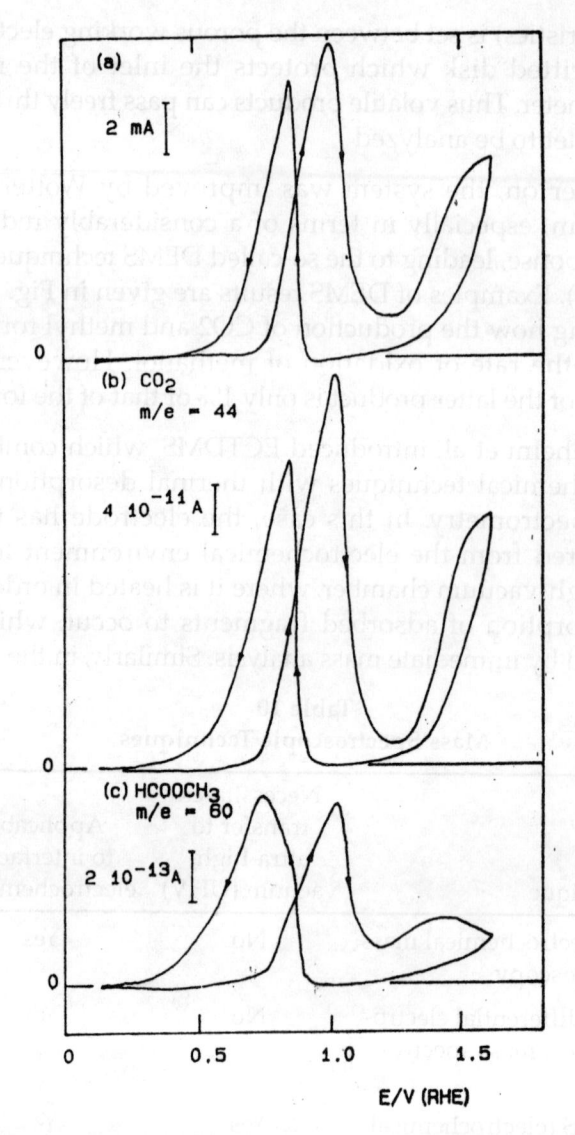

Figure 18. DEMS study of oxidation of 0.1 M CH3Oh at a porous Pt electrode (roughess factor of about 50) in perchloric acid medium. Current (a) ad mass intensity (b and c) signals are shown. Formation of CO_2 (m/e = 44) and methyl formate (m/e = 60) was followed as a function of potential at a potential sweep rate of 20 mVs-1.

technique, the electrode surface, after transfer to the UHV chamber, is bombarded by primary ions with energies on the order of kiloelectron volts. Besides photons and particles, a small quantity of secondary ions are generated and analyzed by mass spectrometry.

As indicated in Table 10, all these techniques have been applied to interfacial electrochemistry, and most of the work is relevant to electrocatalysis. Thus, in the framework of research on fuel cells, investigations of the oxidation of small organic molecules were carried out on platinum electrodes, with the aim of detecting adsorbates and reaction intermediates. The most extensive work was done with DEMS. The adsorption and oxidation of methanol, ethanol, formic acid, reduced CO_2, and propanol have been investigated so far with DEMS, whereas ECTDMS has been used to study methanol and urea. Initially applied to the study of polymer films, the SIMS technique was later used to study methanol adsorption. The potential of the technique for electrochemical applications has been reviewed recently by Trasatti.

(vi) Other in Situ Techniques

Among other in situ techniques available for adsorption studies, those which give the amount of adsorbed species, such as tracer methods and microbalances, are among the most important, since they allow the direct determination of the degree of coverage, 0, and the number of electrons per molecule, N_{epm} [see Section II.3(ii)].

(a) Tracer methods

Among all the techniques which have been used so far for investigations at the electrode/solution interface, radiotracer methods represent certainly the oldest nonspectroscopic approaches. Actually, although they started to be developed as early as 1960, by Blomgren and Bockris, their value in studies of interfacial electrochemistry is still great, and they still compare favorably with the wide range of much, more sophisticated, newly developed spectroscopic techniques.

Based upon the properties of radioactive isotopes, radiotracer techniques provide a direct access to the surface concentration of adsorbed species, which is the most fundamental variable of surface science. Furthermore, in more recent applications, it has been demonstrated that their combination with other experimental techniques (voltammetry, infrared or mass spectroscopies) can be very successful for a better understanding of surface mechanisms.

Basically, tracer methods use trace amounts of an isotope of a particular element (the "tracer") in the same chemical form as the bulk electroactive species. Once the tracer is introduced into the cell, its behavior is followed as a function of time, potential, or any other parameter by means of either radiometry if the isotope is radioactive or of mass spectrometry or IR spectroscopy if the isotope is stable.

A variety of techniques have been developed (Table 11). They are described in detail in several recent extended papers by Horanyi, Kazarinov and Andreev, and Wieckowski, in which all the technical aspects are discussed.

The direct methods, which do not require the removal of the electrode from solution and do not rely on the measurement of analyte concentration, are now preferred to indirect methods, which were not very accurate nor very sensitive.

The principle of the foil method was initially proposed by Frederic Joliot, but the first electrochemical application

Study of the Adsorb Species

Table 11
Tracer Methods

Methods
Indirect methods
Based upon the change of adsorbate concentration in solution and the subsequent change of radiocativity
Based upon the measure of the radioactivity of the electrode surface, after adsorption and its removal solution
Direct methods
Foil method
Thin-layer method
Electrode lowering method

was undoubtedly due to Blomgren and Bockris. I this technique, a gold foil, thin enough for radioactive radiation t cross it, is glued to the external face of a flowing-gas counter. The counter is moved vertically, slowly, toward the solution containing the radiotracer. The radioactivity, which is followed during the approach to the solution surface, increases sharply when the counter comes into contact with the solution, due to adsorption.

The thin-layer method has to be used in conjunction with flow cell techniques. It necessitates that the solution be circulated and allows a fixed position for both the electrode and the radioactive counter. Although interesting, it does not seem to have been greatly developed for electrochemical applications, probably because of the difficulty in maintaining the cleanliness of the solution.

The electrode lowering method was originally proposed by Kazarinov *et al*. The radioactive counter is not in contact with the thin metal electrode, as in the foil method, but is placed below the bottom of the cell, from which it is separated by a thin-film membrane. Once adsorption has taken place, the electrode is lowered, so that only a thin layer of solution remains between the electrode and the membrane, for electrochemical control. A correct estimation of the radioactivity of the surface requires that blank experiments be carried out to measure the various radioactive contributions of the species in the bulk or of those adsorbed onto the membrane. Good sensitivity and accuracy are obtained, and the measurements can be carried out on both smooth and developed surfaces. Isotopes emitting any kind of radiation can be used.

Improvements of the original foil technique have been made by Horanyi and by Sobkowski and Wieckowski, by using metallized plastic foils instead of thin-foil electrodes. Other improvements came as well with the use of plastic or glass scintillators as radioactive detectors, on the surface of which metallic electrodes were vacuum deposited. In the technique developed by Wieckowski, the scintillator is connected to a photomultiplier tube via a light pipe. More recently, combinations of scintillator technology with the electrode-lowering method have been proposed.

The usefulness of radiotracer methods for interfacial electrochemistry is illustrated by 30 years of applications. Most of the studies were done on noble metal electrodes. Besides the adsorption of ions, which has recently been the subject of renewed interest with the coupling of FTIR and tracer methods, the adsorption of a great number of organic compounds was studied. Horanyi[91] demonstrated that organics could be classified into two groups on the basis of their adsorption behavior on platinum. With the first group, the mobility of the adsorbed species is high, so that there is a fast exchange between adsorbed and nonadsorbed molecules. The saturated aliphatic carboxylic acids (apart from formic acid) belong to this group. With the second group, the adsorption is much stronger, leading in most cases to chemisorption. Nearly no exchange occurs with the species in solution. Most alcohols, unsaturated aliphatic acids, and aromatic acids belong to this second group. It should be emphasized that all the so-called "small organic molecules" which are often considered for possible use in fuel cells belong to this second group as well.

As pointed out by Wieckowski, the capability of radiotracer methods to provide information for fuel cell applications has not yet been fully exploited. This is especially true now that improvements in the techniques are making them nearly applicable to well-defined surfaces and single-crystal electrodes. It is also true now that a thin-layer spectroelectrochemical cell fitted with a polished glass scintillator has been successfully designed for application to coupled infrared spectroscopic and radiotracer studies of 14C organic adsorbates.

(b) Isotopic labeling

Although it is not a technique by itself, recourse to the use of isotopic labeling any be of great help for the interpretation of spectra, whether mass, infrared, or Raman. The idea is to induce spectral shifts due to the different mass of the labeled compound as compared to that of the normal molecule. For example, labeled isotopes have been used in conjunction with on-line mass spectroscopy to investigate the adsorbed intermediates resulting from the oxidation of methanol, ethanol, propanol, formic acid, urea, and glucose.

Using infrared spectroscopy, $^{12}CO + {}^{13}CO$ mixtures of various compositions were studied with the aim of evaluating the coupling effects between adsorbates at various coverages. Similarly, the formation of CO in the course of electrooxidation of methanol and formic acid on Pt was studied with FTIR spectroscopy on the basis of $^{12}C/^{13}C$ isotopic substitution.

Depending on the type of study, H_2O, D_2O, or $H_2^{18}O$ may be considered as solvents. If deuterated organics are used, it must be taken into account that during the time scale of the experiment, some of the D atoms do not exchange (those, for instance, which are associated with $-CD_3$, $-CD_2H$, or $-CDH_2$ groups) while others do (such as those of the terminal OD groups of alcohols).

(c) Quartz microbalance

Tracking *in situ* the variations of mass which affect an electrode during electrochemical processes, as a function of potential or as a function of time, has become possible since the development of quartz microbalances sensitive enough to work in the 10^{-9} g range. Briefly, these devices are based upon the properties of a piezoelectric quartz plate, whose resonance frequency varies proportionally with the quantity of a foreign mass added on its surface. Linearity is assumed, provided that the quantity deposited is small. Therefore, using evaporated metals on quartz as electrodes, it is possible to measure changes in their mass during electrochemical processes. The technique has been popularized simultaneously by Bruckenstein and Shay and Kaufman et al. under the name electrochemical quartz crystal microbalance (EQCM). Other variants have been developed more recently; one is based on coupling with a rotating disk electrode, and another is based on a transient analysis of the mass-voltage relationship (Table 12).

Among the applications to electrochemical systems which have been reported so far, those relevant to electrocatalysis concern the upd of metals, the absorption of hydrogen by Pd and the adsorption of ions and polymer films. Work has also been done in the field of corrosion.

However, further development of the technique is expected in the field of electrocatalysis and adsorption studies, since its sensitivity (on the order of a few ns) is sufficient to monitor a full monolayer of adsorbed species (e.g., a monolayer of linearly adsorbed CO would correspond to a mass change of about 60 ng). The main inconvenience, for fundamental studies with single-crystal electrodes, is the need to use evaporated metallic films as electrodes,

Table 12
Techniques Derived from Quartz Microbalances

Technique	Characteristics
QCM (quartz crystal microbalance)	Initially developed for use at the solid/gas interface
EQCM (electrochemical qurtz crystal microbalance)	Designed for *in situ* use at the electrode/solution interface
(rde)-EQCM	Combined rotating disk electorde with EQCM
ac-QEG (ac-quartz electrogravimetry)	Sinusoidal perturbation of current. transient analysis of the mass-voltage relationship is possible

(d) Electron microscopy

Electron microscopy makes use of a thin beam of thermally excited electrons, which are focused on the samle under investigation. In transmission electron microscopy (TEM), the incident electrons penetrate through a thin film of the sample (or a replica of the sample), whereas in scanning electron microscopy (SEM), the incident electrons are scattered by the surface.

Scanning electron microscopy, scanning tunneling microscopy (STM), and their variants (see Table 13) are recent techniques which have seen a large expansion in their use since their ability to examine surfaces on a subnanometer scale was demonstrated. Furthermore, STM can now operate in an electrolytic environment, which makes the technique a new and powerful tool for in situ examination of electrode surfaces.

Some of the methods listed in Table 13 are still under development. As yet, STM is the only one which has been applied to a wide range of electrochemical problems, leading to direct real-space information on the topography of electrode surfaces. Several extensive reviews are now available in the literature.

In STM, a chemically inert fine metal tip is positioned so close to the sample surface (typically a few angstroms) that overlapping of the electron wave functions of the tip and the sample occurs. Thus, when a low voltage is applied

Table 13
In Situ Scanning Electrom Microscopy and Related Techniques

Technique	Applicable to *in situ* studies of the electrode/solution interface
SEM (scanning electron microscopy)	No: operates in UHV
TEM (transmission electron microscopy)	No: operates in UHV
STEM (scanning transmission electron microscopy)	No: operates in UHV
STM (scanning tunneling microscopy)	Yes: can operate in various media including solutions
SECM (scanning electrochemical microscopy)	Yes: can operate in various media including solutions
SICM (scanning ion-conductance microscopy)	Yes: but seems more suitable for for biological applications
AFM (atoic force microscopy)	No?(designed to work on insulating materials)

between the tip and the sample, a very small current flows as a result of electron tunneling. Reference and counter electrodes have to be added for electrochemical applications, in order that the potentials of the tip and the sample are monitored independently. Of great importance are the design of the mechanical setup, especially to avoid vibrations, and the fabrication of the tip.

In the field of electrochemistry, different types of experiments have proved to be successful, since the first report by Dovek *et al* of a specially designed microscope. As predicted by Arvia, topographic images provide direct evidence that electrochemical activation may produce surface roughening, leading to parallel ridges as well as domelike structures. The formation of preferentially oriented electrodes is also possible, depending on the potential programs used.

Underpotential deposition of metals has now been widely investigated by STM. Such studies are highly relevant for electrocatalysis but, probably more interesting with respect to the potential applications, is the discovery that adsorbed molecules could be detected as well. Thus, among other examples, the coadsorption of benzene and carbon monoxide on Rh(111) was recently studied. However, as pointed out by Cataldi *et al.*, the role of adsorbed species in the tunneling process has not yet been clarified, so that one has to be careful in the interpretation of imaging.

Scanning electrochemical microscopy (SECM) is a different approach, developed by Bard and co-workers, that uses an ultramicroelectrode with a tip radius of less than 10 um. Recent applications have demonstrated the feasibility of the technique.

With an increasing number of groups working in the field of scanning microscopy, there is no doubt that new technological improvements of the technique are to be expected. Therefore, more applications to electrocatalysis, especially with respect to fast surface processes, should follow in the near future.

(e) Other experimental approaches

Some other techniques are still in t ; heir infancy with respect to their application to interfacial electrochemistry. It is still not clear whether they have any advantages over the well-established methods described above. However, it must be emphasized that their potentialities have not yet been explored. In this sense, worth mentioning are, for instance, photoacoustic spectroscopy (PAS) and Mössbauer spectroscopy. However, the list is not restrictive.

Although the photoacoustic effect is quite old, the feasibility of its application to electrochemistry has only recently been demonstrated. Either gas-microphone or piezoelectric detectors can be used. Details on basic principles and on cell design may be found in the recent review by Vallet. Up to now, most applications of PAS seem to have concerned the study of oxide films on electrodes.

Mirage spectroscopy, or photothermal deflection spectroscopy (PDS), which is about two orders of magnitude more sensitive than PAS, is able to record both the concentration gradient and the absorption spectrum of species present at the electrode/ electrolyte interface, by monitoring the refractive index gradient produced in the electrolyte by an exciting light (a laser beam perpendicular to the electrode surface). A probe laser beam, parallel to the surface, is deflected by the refractive index gradient, and the amplitude of this deflection as a function of wavelength may provide the absorption spectrum.

Applications of PDS have also mainly concerned the study of oxide layers at copper electrodes or zinc electrodes but a recent study mentions the investigation of (Ir + Ru)/ Ti mixed-oxide electrocatalysts for chlorine evolution.

Mössbauer spectroscopy has been developed for in situ electrochemical experiments and has been recently reviewed. Due to the nature of the technique, applications are restricted to a few metals such as Fe, Co, and Ni. Electrochemically grown films of oxides and of hydroxides, as well as transition-metal macrocycles, used as electrocatalysts for oxygen reduction, have been studied so far.

6

Characterization of the Electrode Material

Since in electrocatalysis both the electrode material and the catalytic surface play a key role in the reaction mechanism and kinetics, particularly for the oxidation of small organic molecules, it is of fundamental importance to use well-controlled electrode materials and to characterize their surface properties.

(i) Role of the Electrode Material

The kinetics of electrocatalytic reactions depend strongly on the chemical nature, the electronic structure, and the geometric texture of the electrode material.

Varying the chemical nature of the electrode material from noble metals to transition metals and sp metals changes the electrocatalytic activity greatly. For example, the oxidation of ethylene glycol in alkaline medium on four different noble metals electrodes (Au, Pd, Pt, Rh) is shown in Fig. 19. Gold gives the highest Current densities, but at too anodic potentials for practical use in fuel cells. Platinum and palladium display moderate electroactivity, whereas rhodium is nearly inactive for alcohol oxidation, due to a strong oxidation of the surface, which blocks the catalytic sites. In acid medium, platinum appears to be the only electrocatalyst sufficiently active for practical applications.

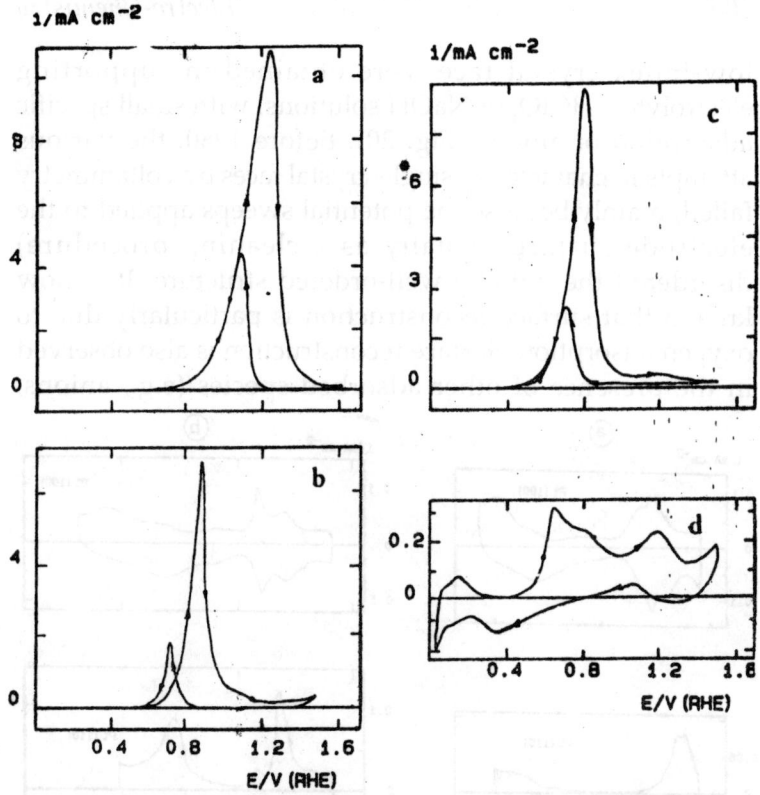

Figure 1. Voltammograms of noble metal electrodes, showing the oxidation of 0.1 M ethylene lycol in alkaline medium (0.1 M NaOH, 250C, 50mVs⁻¹): (a) Au; (b) d; (c) Pt; (d) Rh.

The geometry and the electronic structure of electrode materials is easily varied using different faces of single crystals. Well-defined and well-controlled platinum single-crystal electrodes were first developed and characterized in 1980 by Clavilier et al. They used flame annealing of the electrode surface and then quenched the electrode structure with a droplet of ultrapure water, before transferring the electrode to the electrochemical cell, the electrode surface still being protected from contamination by the water droplet. Under these experimental conditions, well-defined and reproducible voltammograms characteristic of each low-index crystal face were obtained in supporting electrolytes ($HClO_4$ or NaOH solutions) with small specific adsorption of anions (Fig. 20). Before 1980, the various attempts to characterize single-crystal faces by voltammetry failed, mainly because the potential sweeps applied to the electrode surface (usually as a cleaning procedure) disordered the initially well-ordered structure. It is now known that surface reconstruction is particularly due to oxygen adsorption. Surface reconstruction is also observed in the presence of other adsorbed species (e.g., anions,

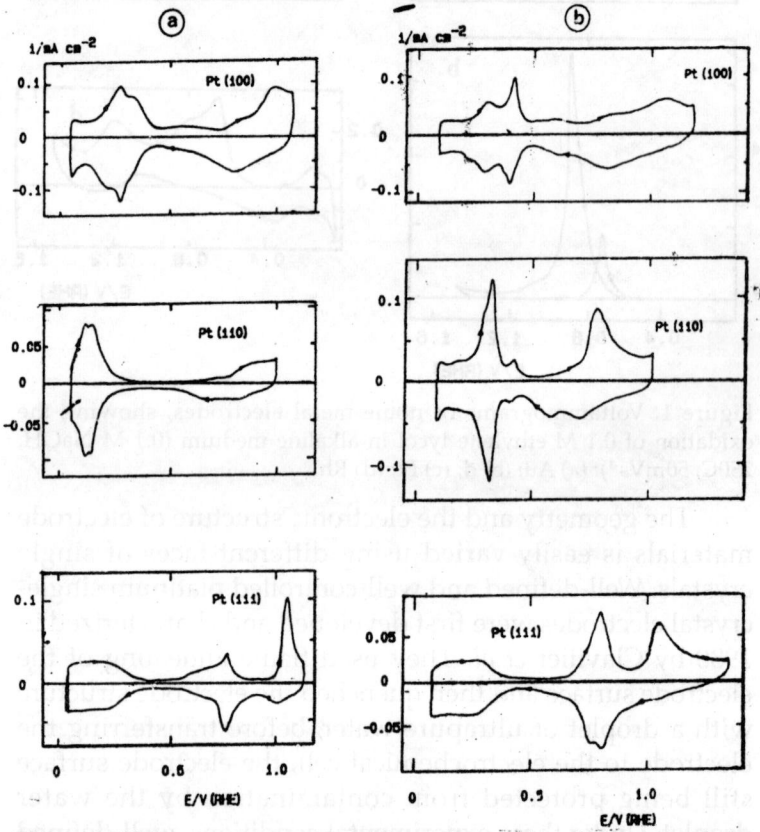

Figure 2. Voltammograms of platinum low-index single crystals in ultrapure supporting electrolytes (25°C, 50mVs⁻¹): (a) 0.1 M HClO4; (b) 0.15 M NaOH.

molecules such as CO), so that it is necessary to perform the flame treatment of the electrode surface after each experiment, to recover the initial surface structure. On the other hand, fast repetitive pulsed potential programs applied to polycrystalline surface have been shown to lead to preferentially oriented surface electrodes, exhibiting similarities to certain low-index faces. Thus, these electrodes behave more or less like single-crystal electrodes in the electrooxidation of various oxygenated species, such as CO and CH_3OH.

Each of the three low-index platinum single-crystal faces, Pt(1 00), Pt(110), and Pt(111), behaves very differently in the electrooxidation of aliphatic oxygenated molecules, such as CO, HCOOH, CH_3OH, C_2H_5OH, C_4H_9OH, glucose, and sorbitol, as discussed in detail in Section III. In the oxidation of small molecules (e.g., CO_2 HCOOH, CH_3OH, C_2H_5OH), the behavior of polycrystalline surface can be easily simulated as the weighted contribution of each single-crystal plane. For very smooth polyoriented platinum surfaces prepared in a similar way to single crystals, the main contribution to the electrocatalytic behavior comes from Pt(110), and the second largest contribution from Pt(100), with a small fraction (about 5%) from Pt(111). Conversely, for larger molecules (e.g., butanol, glucose), it is quite impossible to simulate the electrocatalytic behavior of polycrystalline platinum by that of low-index single-crystal planes, their activity always being much larger (by a factor of 2 to 10) than that of the polycrystalline electrode. This different behavior is probably related to the symmetry and the geometry of the single-crystal surface, which may favor, or may or may not accommodate, the particular structure of the adsorbed species, depending on its size. With small organics, the main adsorbed species is CO, as evidenced by infrared reflectance spectroscopy-see Section III.1-so that the electrocatalytic behavior of the different single-crystal faces is primary determined by CO poisoning. It is therefore observed, in acid medium, that Pt(111), although less active on the first potential sweep, is almost not poisoned at all by CO, so that after a few sweeps it becomes the most active and the most stable face. Conversely, Pt(110), which gives the highest current density on the first sweep, is rapidly poisoned, so that its activity decreases rapidly. The electrocatalytic activity of Pt(100) is more stable with time, although the electrode surface is strongly blocked until potentials of 0.7-0.75 V versus RHE, of which adsorbed CO is oxidized. With larger molecules, such as butanol and glucose, the main adsorbed species is not CO (although a small amount of CO was found by IR reflectance spectroscopy), so that the single-crystal faces are not as sensitive to poisoning, and, consequently, Pt(111) is the most active plane, especially in alkaline medium: it gives a current density of up to 18 mA cm^{-2} for the oxidation of n-butanol, whereas Pt(110) gives 12 mA cm^{-2}, and Pt(100) 3.2 mA cm^{-2} under the same experimental conditions (0. 1 M butanol in 0. 1 M NaOH, 250C, 200 mV s^{-1}), the polycrystalline electrode giving only 2.8 mA cm^{-2} (Ref. 300).

Another way to greatly change the electrocatalytic activity of a given electrode is to modify its surface by underpotential deposition (upd) of various foreign adatoms, such as Ag, As, Bi, Cd, Cu, Ge, Pb, Re, Ru, Sri, and Tl. Depending on their size and on their redox potential, which governs the potential range in which they desorb from the surface by oxidation, as well as on the kind of organic molecule that is oxidized, different enhancement effects or inhibiting effects are found in the literature. One can say roughly that, except for methanol, the more the electrode is sensitive to poisoning, particularly by adsorbed CO, the more important are the enhancement effects of some adatoms. Among the first experimental findings that demonstrated the effect of adatoms was the striking increase of formic acid oxidation currents at a platinum electrode modified by upd of Pb adatoms [see Section III.1(iii.b)]. Another impressive example is the oxidation of ethylene glycol in alkaline medium on modified platinum electrodes. Some of the adatoms considered (Pb, Bi, Tl) greatly increase the current density (by a factor from 6 to 15), leading to values as high as a few hundred milliamperes per square centimeter, whereas others, such as Cd, do not change the current density, but shift the electrode potential cathodically (by about 200 mV), which reduces the oxidation overvoltage [see Section III.3(i)]. In acid medium the behavior of some adatoms is quite different, since, for example, Bi and Tl greatly inhibit the electrooxidation of ethylene glycol.

The interpretation of such effects is not entirely clear, since both geometric and electronic effects modify the catalytic properties of the electrode surface and its ability to adsorb the organic molecule. Many possible explanations have been discussed in the literature-see, for example, the review by Parsons and Vander-Noot—such as the modification of the electronic and redox properties of the surface, the prevention of the formation of blocking species acting as poisoning intermediates, and explanations in terms of the bifunctional theory of electrocatalysis. Moreover, recent experiments using EMIRS clearly showed that, at least in the case of formic acid oxidation in acid medium, at rhodium electrodes, lead adatoms acted by replacing, one by one, bridge-bonded CO, thus reducing the poisoning of the electrode surface. Conversely, Cd adatoms, which are not as strongly adsorbed, were not able to displace adsorbed CO [see Section III.1(iii.c)].

Similarly, the electrocatalytic activity of electrode materials may be improved by using alloys of different kinds, such as binary alloys, plurimetallic alloys, and intermetallic compounds. This approach has been particularly developed for the electrooxidation of hydrogen and the electroreduction of oxygen in fuel cells. For the electrooxidation of small organic molecules, many attempts have also been made to improve the catalytic behavior of the electrode surface by metal alloying. For example, methanol, electrooxidation can be enhanced by using platinum electrodes modified by a second metal [see Section III.1(i.d)). Homogeneous binary alloys, such as Pd/Au, Pt/Pd, and Pt/Rh, which form continuous series of solid solutions, ate particularly interesting since they allow the lattice parameters and the electronic properties of the metal catalyst to be varied in a continuous manner. These metallic alloys usually lead to enhancement of the electrocatalytic oxidation of small organic molecules for a. given alloy composition. Such "synergistic effects" were observed, for example, with the oxidation of formic acid at Pt/Rh Pt/Au, Pt/Pd, and Pd/Au alloy electrodes, with the oxidation of methanol at pt/Rh and pt/pd alloy electrodes, and with the oxidation of ethylene glycol at a Pt/Au alloy electrode. Ternary alloys are even more efficient for the electrocatalytic oxidation of oxygenated molecules, as illustrated by the electrooxidation of ethylene glycol at Pt/Pd/Bi and Pt/Pd/Pb alloy electrodes and at Pt/Au alloy electrodes modified by lead adatoms.

The interpretation of the enhancement of activity is similar to that discussed previously for noble metal electrodes modified by upd of foreign metal adatoms, provided that the surface composition of the catalytic material is similar. The surface composition of metallic alloys is not easy to determine in most cases, as discussed in Section II.4(ii.c). Furthermore, there is generally a preferential dissolution of one metal compound in acid solutions, such as Rh in the case of Pt/ Rh alloys and Pd in the case of Pd/ Au alloys. Even in neutral solutions, it was shown that cyclic voltammetry in the oxygen adsorption region greatly increased the rate of dissolution of one metallic component of the alloy.

In practical systems, such as fuel cells, the catalytic material must be dispersed on a convenient substrate. The first idea was to use smooth foils of the same metal for electrodeposition of the catalytic metal, which gives metallized metal electrodes or metal blacks, platinized platinum being the most popular. Very often, these rough surfaces (the roughness factor of which varies from single-digit values to a few hundred) behave relatively differently from low-surface-area electrocatalysts. For example, the rate of methanol electrooxidation, which is a structure-dependent reaction, does not increase linearly with the true surface area, but, instead, tends to a limiting current intensity, mainly diffusion controlled, when the roughness factor, p, reaches 200. This particular behavior may be related to the change, on the one hand, in the rate-determining step and, on the other hand, in the nature and in the surface distribution of the adsorbed species, as shown by EMIRS.

However, metallized metals and noble metal blacks do not lead to high- surface-area catalysts, since the specific surface area only reaches a few square meters per gram. To achieve a greater dispersion of the electrocatalytic material, it is thus necessary to select high-surface-area electrically conducting substrates, among which graphite, carbon blacks, and activitated carbons are widely used. Specific surface areas, as measured by the method of Brunauer, Emmett, and Teller (BET), vary from a few hundred to a few thousand square meters per gram. Depending on the origin of the carbon substrate, for a given dispersion of platinum or for a given size of the platinum particles, the electrocatalytic properties vary greatly, due to specific metal-support interactions. Ultrahigh dispersions of Pt electrocatalysts, supported on carbon blacks, with specific surface areas of up to 200 m²/g of Pt, corresponding to particle sizes of about 15 Å, display a linear relationship between the mass activity (in A/g) for methanol oxidation and the platinum surface area. The preparation of the electrocatalyst deposit on the carbon surface may be carried out by several methods, including impregnation, ion-exchange adsorption, chemical reduction, thermal decomposition, attack of a Raney metal alloy in base electrolyte, and electrodeposition.

Other supports can also be considered. There is increasing interest in solid polymer electrolytes (SPE), as originally proposed by Hamilton Standard, for the General Electric H_2/O_2 fuel cells, in the framework of the NASA Gemini Space Program, and in perfluorosulfonated membranes (Nafion of Dupont de Nemours and Proton Exchange Membrane of Dow Chemicals), with high ionic conductivity. The use of these has led to the development of extremely powerful systems (a few watts per square centimeter). In recent attempts, the platinum electrocatalyst is directly dispersed at the surface of the SPE membrane, which plays the role of both solid electrolyte and separator. This gives highly efficient systems for water electrolysis and H_2/O_2 fuel cells. These Pt-SPE composites have been also used for the electrocatalytic oxidation of methanol. Binary Pt alloys were also dispersed on SPE membranes, leading again to enhancement effects for methanol electrooxidation, with Sn, Ru, Ir, and Mo.

Another way to achieve very high dispersion of the metal electrocatalyst is by formation in situ of metal microparticles inside a polymer matrix. The recent development of numerous electronic conducting organic polymers, among which polyaniline, polypyrrole, and polythiophene are the most stable in aqueous solutions, offers the possibility of obtaining catalytic electrodes with a very high efficiency. However, apart from studies of hydrogen evolution or oxygen reduction, very few studies concern the electrocatalytic oxidation of organic compounds. One of the first reports, on the oxidation of formic acid at modified platinum polyaniline-coated electrodes, underlined the tenfold increase of the electrocatalytic activity, as compared to that of platinized platinum electrodes, and correspondingly the net decrease in the rate of formation of poisoning species. Comparable behavior was recently found for the oxidation of small organics (HCOOH, CH$_3$OH) at modified platinum polyaniline-, polypyrrole-, polythiophene-, and copolymer (pyrrole-dithiophene)-coated electrodes. Excellent current densities (up to 10 mA cm^{-2}, i.e., about 100 times higher than with smooth platinum) were obtained, with a small amount of platinum (0.1 mg cm^{-2}), and in most cases the forward and backward sweeps were superimposed, showing that strongly chemisorbed poisoning species were not extensively formed on such electrodes.

(ii) Characterization of the Electrocatalyst Surface

To establish precise correlations between the catalytic properties of the electrode material and the kinetics of the electrooxidation of organic compounds, the electrocatalyst surface must be characterized by appropriate methods. This is the only way to conceive and develop new electrocatalysts with high activity for a given reaction, such as the electrooxidation of methanol in a direct methanol fuel cell (DMC).

The real surface area and the surface structure and texture, as well as the surface composition, must be determined, preferably by *in situ* methods. However, some *ex situ* techniques, particularly those developed so far for studying the solid/vacuum interface (electron diffraction and electron spectroscopies), will also be discussed, because of their great interest with regard to the determination of the surface structure and nature of the electrocatalysts. As they require ultra-high vacuum (UHV), their application to electrode surfaces will necessitate the transfer of the electrode from the electrochemical environment to the vacuum chamber. Therefore, the results may be subject to controversy, in that there always remain doubts about whether the double layer and the interface structure are still intact after removal from the solution. However, this is not sufficient reason for rejecting methods that have proven so successful at the solid/vacuum interface. Kolb et al have recently examined this crucial point of the link between electrochemistry and UHV techniques.

(a) Determination of the real surface area

A key physical quantity in . comparing electrocatalytic reactions on electrodes of different nature and structure is the real surface area S_r. The real surface area of the electrocatalyst can be measured by adsorption processes, either physical adsorption involving van der Waals forces or chemisorption leading to the formation of a chemical bond between the catalytic surface and the adsorbate.

Ex situ measurements make use of the adsorption of a gaseous molecule, such as N_2, H_2, O_2, or CO, on the catalyst surface. In the well-known BET method, the volume V_m. of the adsorbed gas (usually N2) corresponding to a full monolayer is first determined, allowing the specific surface area S to be calculated as

$$S = V_m N_A \sigma / V_0 \qquad (40)$$

where V_0 is the molar volume of the gaseous molecule under standard conditions, σ is the cross-sectional area of the adsorbed gas ($\sigma \approx 16.2$ Å^2 for N_2 at -195°C), and N_A is Avogadro's number. The BET method gives relatively reliable S values for high specific surface areas (on the order of a few tens to a few hundreds of square meters per gram).

A variant of the method using radioactive gases, such as Kr-enriched krypton ($\sigma \approx 15.6$ Å^2) is more sensitive, giving surface areas on the order of $0.1 m^2 g^{-1}$.

Gas-phase chemisorption of H_2, CO, or O_2 has been applied to the measurement of surface areas of supported catalysts, such as Pt, Rh, Pd, and Ir, deposited on carbon substrates. The amount of adsorbed gas on the catalyst surface can be determined by static volumetric measurements or, in the continuous-flow method, by gas chromatography after desorption or after titration based on a given surface reaction. For example, preadsorbed oxygen on a Pt surface can be titrated by reaction with either hydrogen or CO at room temperature:

$$Pt-O_{ads} + 3/2 H_2(gas) \rightarrow Pt-H_{ads} + H_2O$$

$$Pt-O_{ads} + 2CO(gas) \rightarrow Pt-CO_{ads} + CO_2$$

These two reactions illustrate the main problem encountered in determining the surface area from the amount of chemisorbed gases, namely, the adsorption stoichiometry. For hydrogen the ratio H/M is usually assumed to be 1 for most metals M. However, for oxygen the O/M ratio can vary between 1 and 2, depending on the catalytic material. Similarly, for chemisorbed CO, the CO/M ratio varies from 1 for linearly bonded CO to 0.5 for bridge-bonded CO, depending on the nature of the catalytic material and on the size of the metallic particles, as discussed in Section II.3(iv).

The gas-phase adsorption methods are not suitable for determining the surface area of smooth electrodes, particularly single crystals, because their sensitivity in measuring the adsorbed volume is too low. Moreover, the adsorption of gases does not necessarily give the real surface area, S_r, of the electrode catalyst, that is, the surface in contact with the electrolyte solution and the electroreactive species. *In situ* measurements of S_r by electrochemical techniques, such as galvanostatic charging curves and potential sweep techniques, are thus much more preferable. Information on both very smooth electrodes (including single crystals) and highly developed electrodes with catalytic particles deposited on high-surface-area substrates can be obtained.

For metals that adsorb hydrogen well (e.g., Pt, Rh, Ir), S_r can be easily determined from the quantity of electricity Q_H^o corresponding to a full monolayer of adsorbed hydrogen ($\theta_H = 1$), assuming that each surface atom is associated with one adsorbed hydrogen atom. Then the theoretical charge per unit surface area (usually per square centimeter), corresponding to deposition or ionization of adsorbed hydrogen, may be calculated provided that the geometry of the surface, that is, the distribution of the surface atoms, is known. This information is only available for single-crystal planes, particularly for the low-index faces. For example, the adsorption of a hydrogen monolayer on the low-index planes of platinum single crystals corresponds to 243 µC for Pt(I 11), 209 µC for Pt(100), and 195 or 149 µC for Pt(110), depending on whether the second row of atoms situated below the upper layer is counted or not. For a polycrystalline electrode, the number of surface atoms will depend on the distribution of the different crystallographic sites, so that the charge per real unit surface associated with the adsorption of hydrogen will be relatively difficult to estimate. For a smooth polycrystalline platinum electrode, a value of 210 µC cm^{-2}, based on the equally weighted contribution of the three low-index planes, is usually assumed, so that the real surface area S_r is calculated as

$$S_r(\text{cm}^2) = \frac{Q_H^o(\mu C)}{210(\mu C/\text{cm}^2)} \qquad (41a)$$

Charasterization of the Electrode Material

From the knowledge of S_r, one may evaluate the roughness factor ρ of the electrode as

$$\rho = S_r/S_E$$

with S_g the geometric area.

For metals that do not adsorb hydrogen (such as gold), the real surface area may be estimated from oxygen adsorption, provided that a full monolayer of adsorbed oxygen can be isolated. This is not so easy, because the charge Q_O associated with the surface oxidation continuously increases with increasing potential before oxygen evolution. However, if the Q_O versus E curve exhibits an inflection point corresponding to a given stoichiometry of the oxygen layer, for example, NO for platinum, it is possible to calculate the real surface area; for example,

$$S_r(\text{cm}^2) = \frac{Q_O(\text{C})}{420(\mu\text{C}/\text{cm}^2)} \quad (41b)$$

where 420 µC cm^{-2} is the charge associated with a full monolayer of oxygen, corresponding to PtO, which requires two electrons per site for its removal.

Instead of adsorbing hydrogen or oxygen on the electrocatalyst surface, and particularly for those metals that do not adsorb them, or that absorb either one or both of them, it is also possible to adsorb foreign metal adatoms by upd. This leads to another way of evaluating S_r, namely, from the quantity of deposited metal, provided that the stoichiometry of the layer, the degree of oxidation of the metal, and its atomic radius are known. For a full monolayer of Cu on Pt, the corresponding quantity of electricity is 417 µC cm^{-2}, consistent with a Cu/Pt stoichiometry of 1 : 1 and two electrons per site [i.e., Cu^{2+} + 2e^- → Cu(0)]. The method is particularly useful in the case of ruthenium electrodes, where hydrogen adsorption, because of competition with hydrogen absorption, or oxygen adsorption, because of the formation of different oxygenated species, cannot be used for surface area measurements. Underpotential deposition of Cu on Ru allows S_r to be evaluated with good accuracy, assuming a Cu/Ru ratio of 1 and a value of 502 µC cm^{-2} for the oxidation of a Cu monolayer. However, the method seems to be limited to S_r measurements for Ru electrodes of small roughness factors ($\rho < 30$), because of the growth of copper multilayers at higher ρ.

The real surface area, S_r, can also be determined from the measurement of the double-layer capacity C_d, assuming that a unit real surface area (1 cm2) corresponds to a known capacity, for example, about 20 µF cm^{-2} for platinum or gold in the double layer, so that

$$S_r(\text{cm}^2) = \frac{C_d(\mu F)}{20(\mu F/\text{cm}^2)} \quad (42)$$

Unfortunately, the reference value is not very well established (values of 16 to 80 µF cm^{-2} have been reported in the literature), and the measured capacity of a rough surface is an underestimate as a result of current line distribution in porous media.

(b) Determination of the surface structure

The catalyst microstructure, or porosity, of porous electrodes is an important factor, which controls both the diffusion and/or adsorption processes and the electron transfer step through the distribution of current and potential lines. The distribution of pore sizes inside the electrode material can be determined from N_2 adsorption isotherms. The method is limited to pore diameters smaller than 300 Å, so that mercury porosimeters have to be used for larger pore diameters. The latter technique presents a major disadvantage, in that after the measurements there remains some Hg, which is known to be an electrocatalytic poison. Moreover, the high pressure necessary for Hg to penetrate into the porous structure may disrupt the pore system.

For some structure-sensitive reactions, the specific activity of the catalyst may depend on the mode of preparation and on the size of the particles. The metallic properties, and thus the electrocatalytic properties, of high-surface-area electrodes depend strongly on the particle size, particularly in the case of small particles.

From the knowledge of the specific surface area S, it is easy to estimate the average particle size, assuming spherical (diameter d) or cubic (dimension d) particles. This gives

$$d = 6/\rho S \qquad (43)$$

where ρ is the density of the metallic particle.

Various physicochemical methods, such as X-ray diffraction (XRD), small-angle X-ray scattering (SAXS), and electron microscopy (EM), allow the particle sizes to be determined.

The presence of small particles in polycrystalline materials causes line broadening of the XRD pattern. The average crystallite size, d, can be evaluated from Scherrer's equation:

$$d = \frac{k\lambda}{\beta \cos \theta} \qquad (44)$$

where k is a constant depending on the crystallite shape (k can vary from 0.7 to 1.7), λ is the wavelength of the X-ray beam, β is the linewidth free of broadening (the natural linewidth), and θ is the Bragg angle [from the Bragg equation 2dhk1 sin θ = $n\lambda$, with d_{hkl} the interplanar spacing in the crystal lattice planes of Miller indices (h, k, l), and n an integer]. This method is useful for crystallite sizes ranging from 50- to 1000-Å average diameter. SAXS allows measurement of smaller particle sizes ranging from 20 to 500 Å. n this technique corrections have to be made for the X-ray scattering intensity of the catalyst support.

The size and shape of electrocatalyst particles can be evaluated from transmission electron microscopy (TEM), by observing electron micrographs. Particle sizes as low as 10 Å can be resolved by EM, but the accuracy is very poor for particles less than 20 Å in diameter. Particle sizes determined from surface area measurements, X-ray diffraction, and electron microscopy are roughly in agreement.

Surface morphology and surface topography are better studied by scanning electron microscopy (SEM) [see Section II.3(vi.d)].

The crystallographic structure of metallic single-crystal electrodes can be conveniently investigated by electron diffraction techniques, such as low-energy electron diffraction (LEED) and reflection high-energy electron diffraction (RHEED). In LEED, the electrons incident on the electrode surface, having an energy typically in the range 10 to 500 eV (which corresponds to a wavelength range of 1200 to 24 Å), can be diffracted by periodic surface structure, that is, by single-crystal planes. This is equivalent to X-ray diffraction by a single-crystal lattice and provides an image of the surface lattice in reciprocal space, from which the surface structure in real space can be derived. The method can also give informations on periodic surface defects (e.g., steps, terraces, long-range disorders), in terms of density, size distribution, etc.

LEED and RHEED have provided very useful information on the reconstruction of the surface of noble metal single-crystal electrodes, which results from adsorption-desorption processes of the oxygen layers in cyclic voltammetry.

LEED also provides useful information on the adsorption of different atoms and molecules on well-defined surfaces, such as single-crystal planes.

Finally, Rutherford backscattering spectroscopy (RBS) is also an ex situ technique which can be applied for electrode surface analysis, as reviewed by Kötz. Although most of the work was done on oxides and oxide formation, the RBS technique has also been applied by Hyde et al. to characterize the distribution of Pt particles in a fuel cell catalyst.

It is also worth mentioning that RBS can be performed *in situ* using a thin silicon membrane, on which a metallic film electrode is evaporated and through which the accelerated ionic beam passes, reaching the electrode/electrolyte interface before being backscattered. Such an arrangement allowed Kötz et al. to study Cu deposition on Ir electrodes.

The present list of techniques is only a brief introduction to the possibilities offered by *ex situ* UHV spectroscopies. It is not intended to cover all the aspects nor all the approaches to problems connected with interfacial electrochemistry. It was just presented in order to show the profit that electrochemists may draw from the use of powerful techniques that have long been developed for surface chemical investigations of the metal/vacuum interface, as recently pointed out by Bockris and González-Martin. Their main application in the field of electrocatalysis is the control and characterization of well-defined electrode surfaces, such as single-crystal electrodes and preferentially oriented surfaces.

(c) Determination of the surface composition of multicomponent catalysts

The surface composition of many alloys and bi- or plurimetallic electrodes is different from that of the bulk material, because of preferential dissolution of one component in an aggressive electrolytic solution (acid electrolyte) and/or the influence of adsorbed species (adsorbed oxygen resulting from potential cycling in base electrolyte or adsorbed electroreactive species such as CO).

X-ray diffraction and microprobe electron microscopy (analytical TEM, which makes use of the energy of the X rays generated by electron bombardment to identify elements with atomic number $Z > 10$) can be used to determine the surface composition. XRD gives the composition of binary alloys from an accurate measurement of the lattice parameter, which is, to a first approximation, a linear combination of the lattice parameters of the pure metals. This is the case for homogeneous alloys, such as Pt/Rh alloys. However, the determined composition is characteristic of a surface layer a few microns thick, which is not very significant for electrocatalysis, since only the first atomic layers are electroactive.

Electron spectroscopies, such as ultraviolet photoelectron spectroscopy (UPS), X-ray photoelectron spectroscopy (XPS), also known as electron spectroscopy for chemical analysis (ESCA), and Auger electron spectroscopy (AES) are very powerful techniques for determining the surface composition of metallic electrodes. Since the penetration of electrons (on the order of a few tens of angstroms, corresponding to two to five atomic layers) is much lower than that of X rays, electron spectroscopy gives information much more relevant to the surface composition. Electron spectroscopy can give qualitative and quantitative information about the surface from the position in energy and the intensity of the spectral line characteristic of the electron levels. Because the position of the line is strongly dependent on the environment of the atom (chemical shift), it can be used to determine the oxidation state of the element. For example, Pt(II) species have been identified by ESCA as one valence state of platinum dispersed on a carbon support that is active for the electrocatalytic oxidation of methanol. XPS has also been used to analyze polymer films and to monitor the surface concentration of anions and cations on gold emerged electrodes.

AES is much more sensitive than ESCA, and surface quantities as small as 10^{-3} of a monolayer can be monitored. However, quantitative analysis by AES is much more delicate, since numerous Auger peaks may overlap and are difficult to separate. AES can be performed in the same UHV chamber as LEED, and both methods are usually simultaneously available. Therefore, structural information on the investigated surface can be obtained under the same experimental conditions.

Similarly, AES can be coupled with electrochemical experiments, as initially shown by Ishikawa and Hubbard and Revie et al. As pointed out in recent reviews by Aberdam and Durand et al. AES can provide detailed information on the upd of metals, which is relevant to electrocatalysis.

Combining these different techniques with cyclic voltammetry and energy loss electron spectroscopy (EELS) enabled Batina et al. to demonstrate the utility of ex situ techniques for electrochemistry. These surface spectroscopies in UHV were used to study the adsorption on Pt single crystals from aqueous solutions of a wide series of organic compounds of biological or pharmacological interest.

As stated before, most of the techniques described above require the transfer of the electrode surface to the

UHV chamber of the apparatus, which always raises the question of the validity of measurements in a nonelectrochemical environment (with thus no potential control of the electrode surface, possible change of the surface properties in vacuum, risk of contamination of the surface during the transfer operation, etc.). In situ methods are thus much more preferable.

The only method described in the literature to evaluate in situ the surface composition of metal alloys is the voltammetric determination of the peak potential, E_p, associated with oxygen desorption during the negative sweep of a voltammogram recorded in the pure supporting electrolyte. For a homogeneous alloy (e.g., Pt/Rh, Au/Pd), there is only one desorption peak, the position of which is located between those of the pure metals and is, to a first approximation, a linear function of the surface composition. A good correlation was found between such in situ determinations and *ex situ* results obtained with ESCA. For a heterogeneous alloy (such as Au/Pt), the number of oxygen desorption peaks gives the number of different surface phases, and the quantity of oxygen adsorbed on each phase gives its surface area, so that the surface composition can be determined.

Role of the Electrolytic Solution

The composition of the aqueous electrolytic solution is also of great importance, because it can influence greatly the behavior of the electrocatalytic material, and it may modify drastically the oxidation kinetics of small organic molecules, through pH effects and interaction effects between adsorbed species.

(i) Role of the Water Adsorption Residues

Water molecules and their adsorbed residues (H_{ads},

OH_{ads}, O_{ads} play a key role in the oxidation mechanism, particularly in the case of the oxidation of alcohols to carboxylic acids (or to carbon dioxide), which requires one extra oxygen atom to be supplied:

$$R-CH_2OH + H_2O \rightarrow R-COOH + 4H^+ + 4e^-$$

This oxygen atom may come either from an adsorbed water molecule, $(H_2O)ads$, or more probably from adsorbed hydroxyl, OH_{ads}, or adsorbed oxygen, O_{ads}, depending on the electrode potential and the pH of the solution.

By varying the potential limit of a voltammetric sweep, it is very easy to control the amount of adsorbed hydrogen

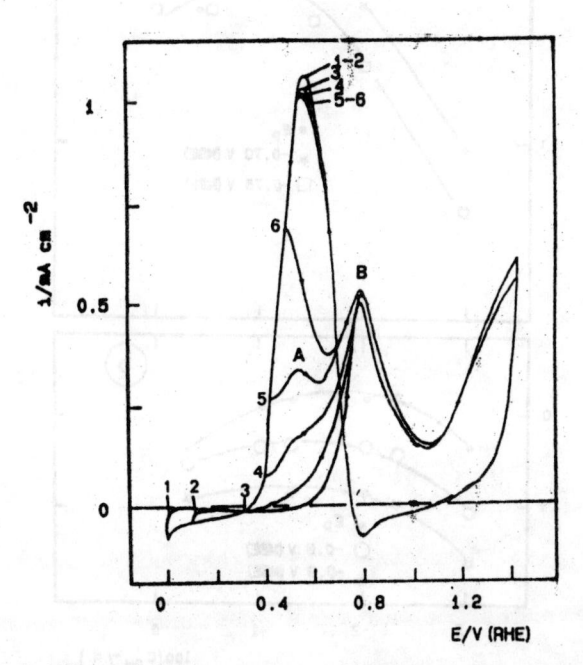

Figure 3. Effect of the lower potential limit on the voltammogram showing the oxidation of 0.1 M ethylene glycol in 0.5 M H_2SO_4 at a platinum electrode (25°C, 50 mVs^{-1}). Curves 1-6 are obtained by increasing the lower potential limit. A and B represent two different two different oxidation peaks during the positive sweep.

or of adsorbed hydroxyl or adsorbed oxygen at the electrode surface. For example, the effect of adsorbed hydrogen on the electrooxidation of ethylene glycol at a platinum electrode in sulfuric acid can be easily seen in Fig.

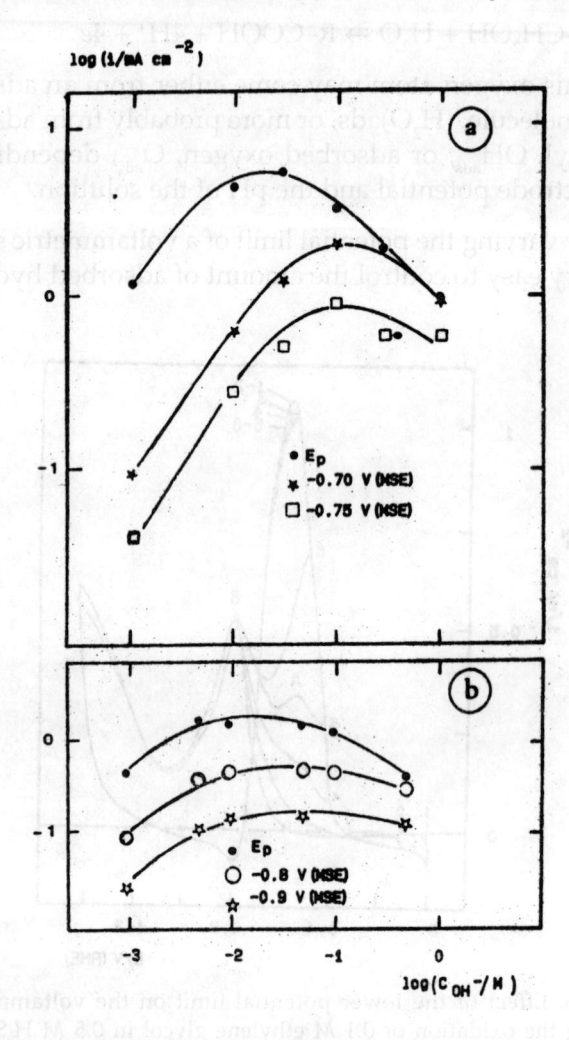

Figure 4. Effect of the hydroxyl ion concentration on the current densities of the oxidation Of 0-1 M CH3OH (a) and 0.1 M n-butanol (b) in 0.1 M NaOH at a platinum electrode (25*C, 50 mV s').

21. When the lower potential limit Ec is made more negative, the current at the first oxidation peak decreases drastically. This could be due to the presence of a poisoning species formed between the adsorbed organic residue and adsorbed hydrogen, as was checked by modifying the electrocatalytic surface by upd of cadmium adatoms.[18,306] In fact, it is now known, from voltammetric and EMIRS measurements, that cadmium adatoms are not able to displace adsorbed CO. However, they still can prevent hydrogen adsorption, and, with this modification of the surface, there was no longer an effect of the lower potential limit on the first oxidation peak A, which was as intense as the second oxidation peak B.

The effects of OH_{ads} and O_{ads} are more subtle, since, on the one hand, the oxidation reaction needs such species in order to go to completion and, on the other hand, the presence of relatively strongly adsorbed oxygenated species may block some electrode sites, preventing adsorption of the organic molecule. Therefore, the oxidation current passes through a maximum when OOH is varied, either by changing the electrode potential or the pH of the solution. For example, the oxidation current, at a given electrode potential, of methanol or n-butanol gives a volcano-shaped curve when plotted as a function of the bulk concentration of hydroxyl ions (Fig. 22). The OH^- concentration at the maximum will depend on the nature and concentration of the alcohol. In the case of methanol oxidation, the maximum corresponds to an equimolar ratio of the alcohol and hydroxyl ions.

(ii) Role of the Anions of the Supporting Electrolyte

Most anions of the supporting electrolyte are specifically adsorbed at the electrode surface and thus compete with the organic molecule for adsorption sites. If anions, such as I^-, Br^-, and Cl^-, are so strongly adsorbed that

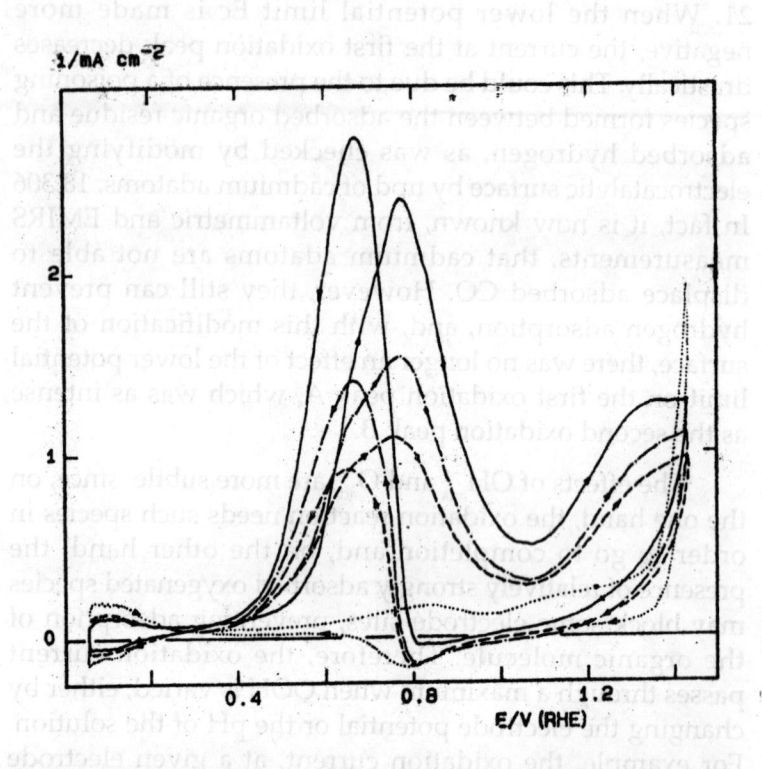

Figure 5. Effect of chloride ions on the voltammograms showing the oxidation of 0.1 M CH$_3$OH in 0.5 M H$_2$SO$_4$ at a platinum electrode (25°C, 50 MVs^{-1}). Concentration of Cl$^-$: -----, 0; ----. ---, 10^{-5} - - -, 10^{-4}; . . ., 10^{-3} M.

they can prevent adsorption of the organic molecule, the oxidation current becomes very small and may even disappear. For example, the effect of chloride ions on the electrooxidation of methanol is illustrated in Fig. 23. Hydroxyl ions and, to a smaller extent, perchlorate ions are not specifically adsorbed at the electrode surface, while sulfate, bisulfate, and phosphate ions are much more strongly adsorbed. For example, the adsorption of sulfate and bisulfate ions on Pt(111) is so strong that it prevents adsorption of methanol, and the oxidation currents are drastically reduced by comparison to those obtained in perchloric acid medium, which are ten times higher (Fig. 24).

Other acid electrolytes, such as trifluoromethanesulfonic acid, CF_3SO_3H have been also considered as supporting electrolytes in a direct methanol fuel cell. However, although less adsorbed than sulfate or bisulfate ions, $CF_3SO_3^-$ results in lower activity of platinum electrocatalysts for methanol oxidation, particularly at elevated temperatures (60 to 800C. This is probably due to decomposition of the electrolyte, leading to adsorbed sulfur species, which poison the catalytic surface and cause the oxidation current to decrease.

In alkaline solutions, hydroxyl ions are not strongly adsorbed at most noble metal electrodes. Moreover, they can compete efficiently for adsorption sites with CO

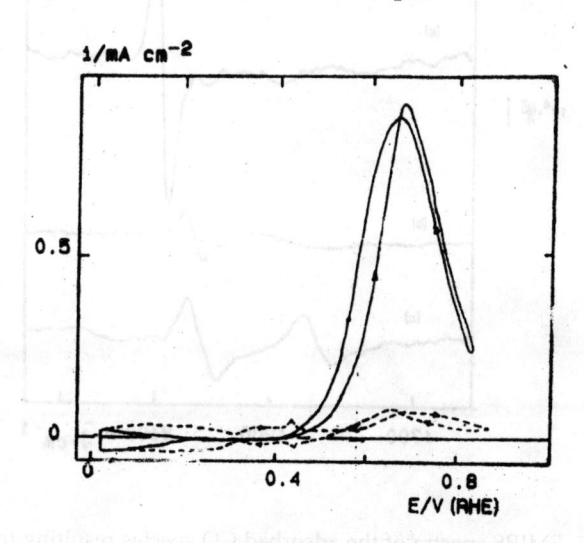

Figure 6. Voltammograms showing the oxidation of 0.1 M CH_3OH at a Pt(111) single-crystal electrode (25⁰C, 50 mVs⁻¹): ─────, 0.1 M $HClO_4$; - - -, 0.5 M H_2SO_4.

poisoning species, so that the amount of adsorbed CO is reduced and higher current densities for the oxidation of alcohols are obtained in alkaline medium. The smaller extent of poisoning phenomena is displayed in the voltammograms, where the forward and backward sweeps are quite well superimposed in the potential range of alcohol oxidation, without blockage of the active sites by OH_{ads} or O_{ads}. This is also evidenced in the EMIRS spectra of the CO poisoning species, whose intensity is decreased when the pH of the solution is increased; see, for example, the EMIRS spectra obtained with ethylene glycol as a function of pH (Fig. 25).

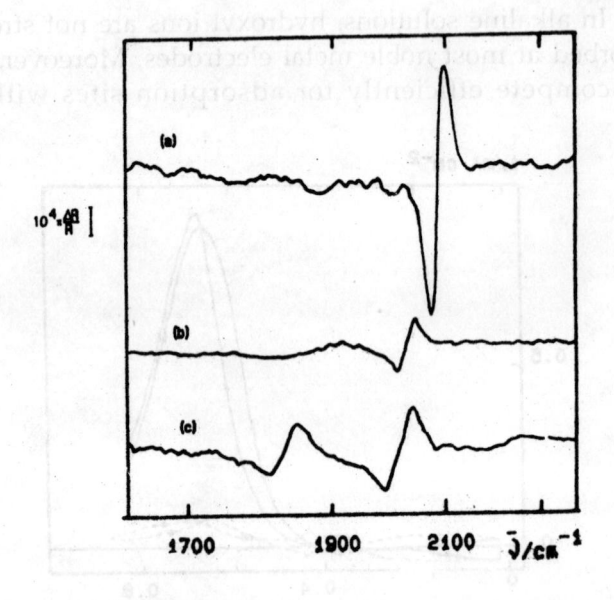

Figure 7. EMIRS spectra of the adsorbed CO species resulting from the dissociative chemisorption of ethylene glycol on a platinum electrode at room temperature as a function of pH: (a) pH ≈ 1; (b) pH ≈ 5; (c) pH ≈ 13.

SELECTED EXAMPLES

Most of the examples selected in the following pages will deal with the electrocatalytic oxidation of relatively small oxygenated organic molecules soluble in aqueous electrolytic solutions. They concern particularly the oxidation of alcohols and of the derivatives of their oxidation, from one to six carbon atoms in length, at noble metal electrodes (e.g., Pt, Rh, Pd, Au).

Oxidation of One-Carbon Molecules

Simple one-carbon-atom organic molecules such as methanol, formaldehyde, and formic acid have several advantages with respect to their use in fuel cells. Their structures are simple enough that one can hope for a complete understanding of the reaction mechanism. In addition, the high energy densities of these compounds, their ease of handling and storage, and their availability from biomass are particularly attractive.

Unfortunately, these fuels give only low current densities except on platinum electrodes. Moreover, mainly on platinum, poisoning effects lead to a decrease of the initial performance. It was recently recognized, for example, from IR reflectance spectroscopic studies, that adsorbed CO plays a key role in the poisoning phenomenon, so that the electrocatalytic oxidation of dissolved CO will also be discussed in this section. From a fundamental point of view, there have been numerous studies on the electrooxidation of these simple compounds during the past 15 years, which have led to the proposal of new ideas and hypotheses for understanding the reaction mechanisms.

(i) Methanol

Methanol is one of the best candidates as a fuel in fuel cell systems for terrestrial applications.

In a recent paper, Parsons and VanderNoot reviewed the electrocatalytic oxidation of small organic molecules and pointed out that more than 40 papers were devoted to the electrooxidation of methanol between 1981 and 1987. According to these authors, a distinction should be made between work done before and after 1980. Various reviews on work published before 1980 are available. Before 1980, although many groups worked on this subject, the mechanism of methanol oxidation was not elucidated, so that two kinds of mechanisms were proposed in the literature, depending on the assumed nature of the adsorbed intermediates responsible for the poisoning effect. Until 1980, no definitive experimental data were available that could help in making a clear choice between the two possibilities. Since 1980, different research groups have considered the whole problem again, taking into account the surface structure of the electrode, but, above all, using more rigorous, reproducible, and well-controlled experimental conditions. The newly available in situ spectroscopic techniques and on-line analytical methods have allowed a detailed study of the adsorbed species (reactive and poisoning species) and of the reaction intermediates, making obsolete many of the older works, as well as the reaction mechanisms based only on electrochemical measurements.

In acid medium the overall reaction of methanol electrooxidation can be written as follows:

$$CH_3OH + H_2O \rightarrow CO_2 + 6H^+ + 6e^-$$

with a standard electrode potential $E^0 = 0.02$ V versus SHE, as calculated from thermodynamic data.

This reaction involves several adsorption steps, including the formation of chemisorbed residues, leading to a decrease in the catalytic activity of the electrode surface.

Charasterization of the Electrode Material

In fact, the general mechanism can be summarized as follows:

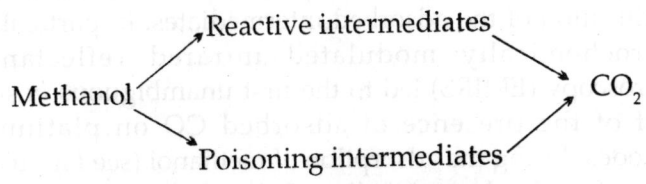

```
             Reactive intermediates
            ↗                       ↘
  Methanol                              CO₂
            ↘                       ↗
             Poisoning intermediates
```

where the reactive intermediates are weakly adsorbed on the electrode surface, in contrast to the poisoning intermediates, which are strongly bonded.

(a) Identification of the adsorbed species by spectroelectrochemical techniques

The nature of the adsorbed intermediates responsible for electrode poisoning led, over a long period, to polemic discussions. According to Bagotsky and Vassiliev, the poisoning intermediate could be COH. This suggestion was the result of assumptions made on the basis of electrochemical measurements, such as the determination of the number of electrons involved in the oxidation of the adsorbed intermediate [N_{epm}; see Section II.3(ii)]. However, other groups proposed CO as the adsorbed intermediate, using the same type of experiments.

This discrepancy in the interpretation of the same type of experiments could be attributed to a lack of accuracy of the measurements, especially because the electrode structure was not well controlled. Until the early eighties, no experimental methods were really suitable for the *in situ* identification of adsorbed intermediates. Only UHV techniques were available, but they provided information only after transfer of the electrode from the electrochemical cell to the vacuum chamber, which was not easy to do under controlled electrochemical conditions.

The development of *in situ* infrared spectroscopic techniques allowed significant progress to be made in the identification of the adsorbed intermediates. In particular, electrochemically modulated infrared reflectance spectroscopy (EMIRS) led to the first unambiguous *in situ* proof of the presence of adsorbed CO on platinum electrodes during the adsorption of methanol (see Fig. 16a). CO was found to be mainly linearly bonded to the electrode surface (IR absorption band at around 2060 cm^{-1}), but bridge-bonded species were also detected (very small band at around 1850-1900 cm^{-1}). These two adsorbed CO species are responsible for the poisoning phenomena observed during working of a direct methanol fuel cell.

Such results were widely confirmed by several groups, not only by EMIRS, but also by similar methods, such as SNIFTIRS and IRRAS, which are based upon the use of Fourier transform IR spectrometers [see Section II.3(iv.e)].

However, these early conclusions were partially refuted by other observations made using different methods based on electrochemical measurements coupled with mass spectroscopy, such as differential electrochemical mass spectroscopy (DEMS) and electrochemical thermal desorption mass spectroscopy (ECTDMS) [see Section II.3(v.e)]. These techniques, which require large-area electrodes made from the catalytic material (platinum lacquer) mixed with Teflon powder, allowed the detection of adsorbed CO, but the main species identified were claimed to be either adsorbed COH or adsorbed CHO.

These apparently contradictory observations have to be related to the different experimental conditions used, particularly the electrode structures, which were different in the two sets of experiments. Further work made clear that the chemical nature and the surface distribution of adsorbed species depend on the electrode structure and on its degree of coverage by the adsorbed species. This was confirmed recently by an EMIRS study, in which the methanol concentration was varied over a wide range. It was demonstrated that different adsorbed species exist simultaneously on smooth polycrystalline platinum at low coverages, that is, at low concentrations of methanol in solution (Fig. 26). Under these conditions, it is reasonable to think that adsorbed CO and other adsorbed species such as formyl, CHO_{ads} (complex band at around 1700 cm^{-1}), exist simultaneously on the electrode surface. It follows that these results, obtained by EMIRS, could be correlated to those obtained by mass spectroscopy.

The coexistence of poisoning species (CO_{ads}) and reactive species (CHO_{ads}) on the electrode surface during the electrooxidation of methanol is the first conclusion of these in situ studies.

(b) Correlation between electrochemical data and in situ spectroscopic measurements

As seen above, the species responsible for poisoning is identified as adsorbed CO, mainly linearly bonded. This conclusion can be well correlated to pure electrochemical measurements. When a species is adsorbed at a constant potential during a given adsorption time, it becomes possible to calculate the electrode coverage and the number of electrons per site, N_{eps}, from the quantities of electricity associated with its oxidation and with the deposition or oxidation of adsorbed hydrogen [see Section II.3(ii)].

The technique used to obtain such results, potential programmed voltammetry (PPV), is illustrated in Fig. 27. The electrode surface is submitted to a potential program including a preparation procedure, an adsorption plateau (of duration t_{ads} at E_{ads} followed by a fast negative sweep to deposit adsorbed hydrogen on the free electrode sites, and a fast positive sweep to oxidize the species previously adsorbed. The sweep rate has to be chosen carefully: it must

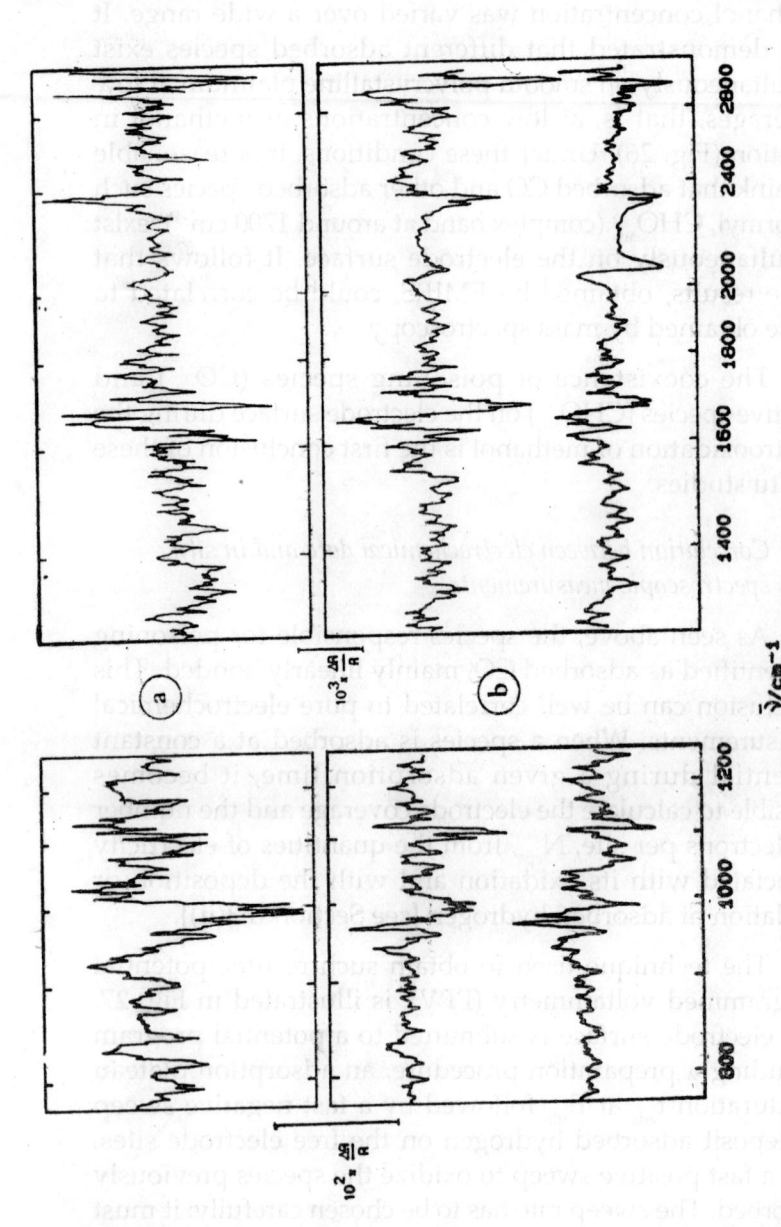

Figure 8. EMIRS spectra of the species resulting from adsorption of methanol (5×10^{-3}) M in 0.5 M $HClO_4$ on a polycrystalline Pt electrode for different times of accumulation $\Delta E = 0.4$ V; $\bar{E} = 0.2$ V versus RHE; room temperature): (a) 1st scan; (b) 10th and 25th scans.

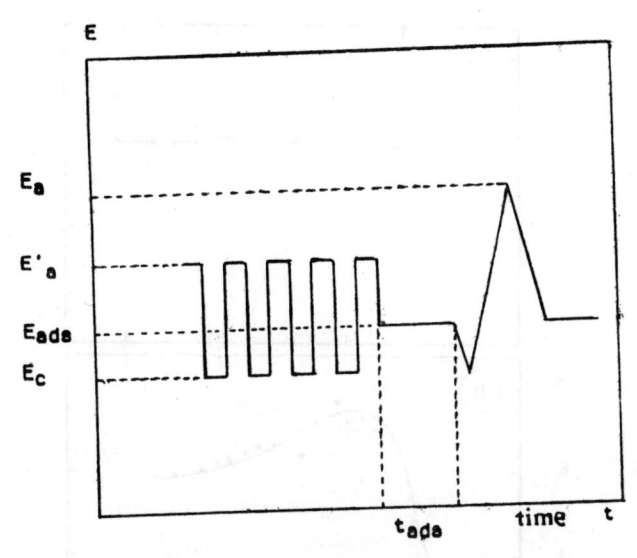

Figure 9. Potential program for PPV measurements.

be sufficiently high to ensure negligible readsorption of methanol and sufficiently low to obtain good voltammograms without distortion. The different quantities of electricity involved- Q_{org}, associated with the oxidation of adsorbed species, and Q_H^o, and $Q_{H'}$ associated with the desorption of hydrogen without and with organic species in the solution, respectively- are calculated by integration of the voltammetric peaks. The number of electrons per site, N_{eps}, and the degree of electrode coverage by the adsorbed organic, Θ_{org}, are then deduced using Eqs. (11) and (9) given in Section 11.3(ii). Θ_{org} is in fact the fraction of the electrode surface not covered by adsorbed hydrogen. It should be noted that these different quantities of electricity must be estimated very cautiously and that several corrections, such as the change of the double-layer capacity in the presence of the adsorbed species, need to be made very carefully.

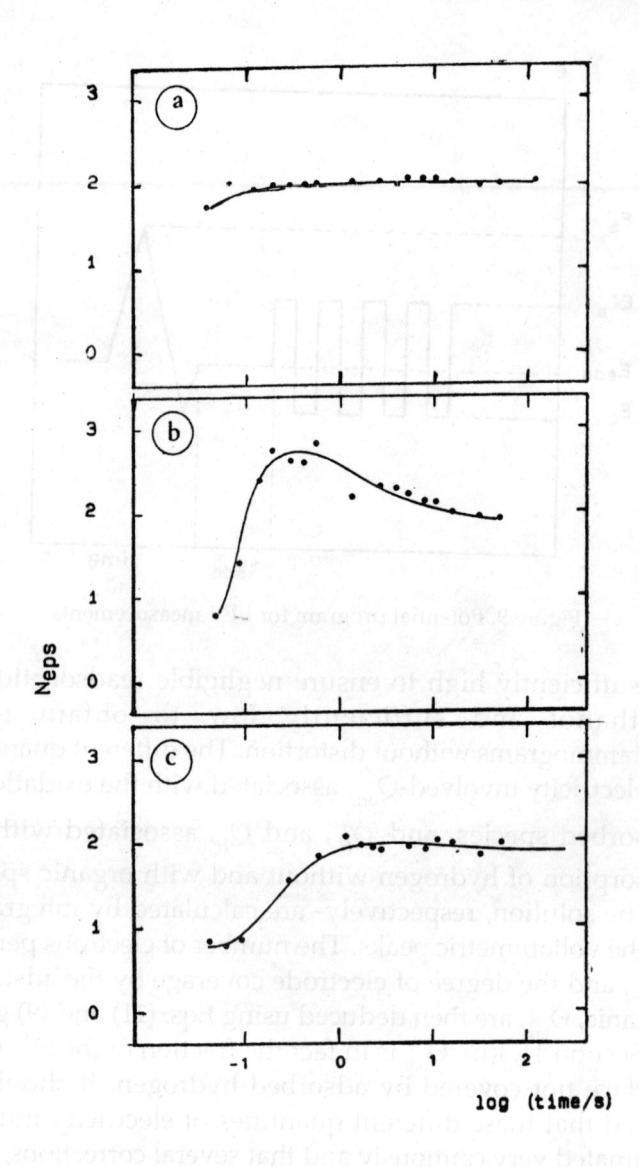

Figure 10. Number of electron per site, Neps, versus the logarithm of adsorption time for: (a) 0.1 M CH_3OH at polycrystalline Pt; (b) 0.001 M CH_3OH at polycrystalline Pt; (c) 0.1 M CH_3OH at a Pt(100) single crystal.

A typical example is shown in Fig. 28 for three different cases involving two structures [polycrystalline platinum and Pt(100)] and with two different concentrations of methanol. In the case of polycrystalline platinum with 0.1 M methanol, N_{eps} is always close to 2, irrespective of the adsorption time (Fig. 28a). With lower methanol concentrations, N_{eps} reaches 2 only for adsorption times longer than 2 s. For $t_{ads} < 2$ s, N_{eps} is in the range 2.5-2.7, and for very short adsorption time, $N_{eps} \approx 1$ (Fig. 28b).

On Pt(100), with 0.1 M methanol, $N_{eps} \approx 1$ for short adsorption times and increases abruptly for $t_{ads} > 0.5$ s, again reaching the value of 2 (Fig. 28c).

These results can be interpreted rather easily by looking at the possible adsorbed species formed from methanol dissociation and the corresponding N_{eps} [see Table 4 in Section II.3(ii)].

A N_{eps} of 2 can correspond only to linearly bonded CO, so that, at $t_{ads} > 2$ s, in all cases, the main species present on the electrode surface is linearly bonded CO (CO_L). When the concentration of methanol is low, N_{eps} can reach a value close to 3 for polycrystalline electrodes, which corresponds only to CHO_{ads}. A N_{eps} of 1 or between 1 and 2 corresponds probably to a mixture of different species including bridge-bonded CO (CO_B).

All these results fit perfectly well with the conclusions drawn from EMIRS measurements. CO_L is the main poisoning species, and CHO the reactive intermediate. On Pt(100), the EMIRS results showed always a mixture of linearly and bridge-bonded CO with a clear band corresponding to CHO at short adsorption times.

Some attempts were made to avoid electrode poisoning by periodic removal of the adsorbed CO species, using a pulse potential program applied to the platinum electrode.

(c) Use of model surfaces (single-crystal electrodes)

Rigorous control of the electrode surface before the experiments and especially during the transfer procedure led to significant progress in the understanding of the reaction mechanism of electrooxidation of methanol. For fundamental studies, single-crystal electrodes are ideal model systems. However, the first attempts to observe possible effects of superficial structure in electrocatalysis on platinum single crystals failed, probably because of the difficulty of transferring the electrode, after its preparation, into the electrolyte without contamination of its surface. Furthermore, the electrochemical pretreatments, which were supposed to "clean the surface," in fact led to drastic modifications and reconstruction of the surface structure.

A new and simple method was proposed by Clavilier et al. in 1980, in order to perform such experiments with single-crystal platinum electrodes, well oriented and uncontaminated during transfer to the electrochemical cell. It was thus possible to observe drastic structural effects on the adsorption- desorption processes of both hydrogen and oxygen at low-index single-crystal platinum electrodes in contact with the supporting electrolyte alone (see Fig. 20).

The electrocatalytic oxidation of methanol is also structure dependent (Fig. 29a). Pt(110) is the most active plane, but also the most sensitive to poisoning, which leads to a rapid decrease in the current densities observed. The Pt(111) plane appears to be less sensitive to poisoning phenomena, even though the current densities are rather weak. Finally, the Pt(100) plane is totally blocked over a large range of potentials, but the current increases sharply once the adsorbed blocking species are removed at higher potentials, the maximum current densities remaining very stable with time.

Confirmation of these structural effects was obtained from EMIRS experiments using platinum single-crystal electrodes (Fig. 29b). The reactive intermediate (CHO_{ads}) was clearly observed in the EMIRS spectrum in the case of Pt(1 11) and Pt(1 00) electrodes, while adsorbed CO was observed for all three surface orientations studied. However, linearly bonded CO_{ads} was the only species detected on Pt (110), whereas two kinds of CO (linearly and bridge-bonded) were clearly present on Pt (100). The latter observation is very interesting, because it is an indication that the blocking phenomena observed with the Pt(100) plane could result from lateral interactions between these two kinds of adsorbed CO species.

(d) Modification of the electrode surface composition

The previous examples illustrated the case of methanol oxidation at platinum electrodes. In acid medium, platinum appears to be the most efficient catalyst for the electrooxidation of methanol even though a few other noble metals, such as palladium and rhodium, also exhibit activity, albeit very low. In alkaline medium, the problem is somewhat different, and while platinum appears to be the best catalyst, significant activities are obtained with palladium and rhodium electrodes and, to a smaller extent, with gold electrode.

Different attempts have been made, through the use of alloys and upd-modified electrodes, to increase the electrocatalytic activity and to decrease the extent of poisoning phenomena by modifying the nature of the surface.

In the case of alloys, different bimetallic electrocatalysts have been investigated, including Pt/Sn, Pt/ Rh, Pt/ Pd, and Pt/ Ru. Only Pt/Ru, and perhaps Pt/ Sn, electrodes display a greater activity than pure platinum. The Pt/Ru alloys appear to be the most promising, with a

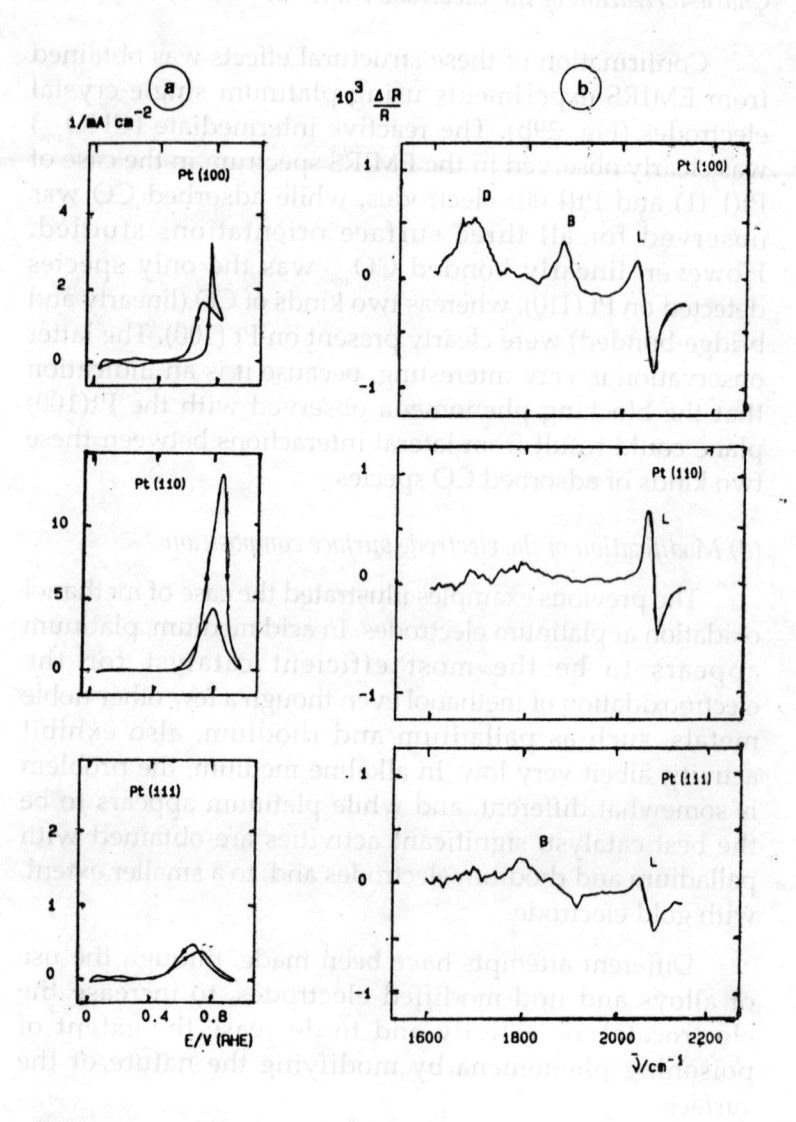

Figure 11. Adsorption and oxidation of 0.1 M CH₃OH in 0.5 M HClO₄ at the three low-index faces of platinum single crystals (room temperature): (a) voltammograms (50 mVs-1; first sweep); (b) EMIRs spectra (ΔE = 0.4V; $\overline{E}$ 0.35 V versus RHE)

negative shift of the polarization curves compared to those obtained with pure platinum and a decrease in., poisoning. However, surface enrichment in one metal could arise under working conditions, thus modifying the activity of the electrodes.

The second possibility is to modify the properties of the electrocatalyst by adsorption of a foreign metal on its surface by upd. Several review papers have been published on this subject. The atoms adsorbed on the electrode surface, the so-called adatoms, modify greatly the adsorption of organics, leading generally to positive effects with significant enhancements of the electrocatalytic activity of platinum. However, methanol seems to be a particular case in which the influence of adatoms is very limited. Only Pb, Ru, Bi , Sn, and Mo adatoms increase slightly the activity of platinum in the electrooxidation of methanol, and only at low potentials.

(e) Increase of the real surface area

For practical applications, the key point is to decrease the amount of platinum used. it is necessary to design electrodes with a large active area but with the smallest amount possible of precious metals. The electrode has generally been made using a conducting support, such as carbon, with deposition of small platinum particles, generally by reduction of a platinum salt.

Numerous studies have been carried out to examine possible effects of the particle size. For example, systematic studies were made in the case of the electroreduction of oxygen and the adsorption of hydrogen; no significant size effects were observed, except for O_2 reduction. Recently, two different groups investigated this problem for the case of electrooxidation of methanol in acid medium. According to Goodenough *et al.*, the most efficient electrodes are those which contain both small crystallites (diameter ≈ 20 Å) and a minimum amount of ionic Pt species. Watanabe *et al.* disagreed with these conclusions. They observed no effects of platinum crystallite size, even at high platinum dispersion, such as 70% (corresponding to particles as small as 14 Å). They concluded that increasing the dispersion of platinum on carbon is an interesting possibility for achieving higher catalytic activity with low amounts of platinum.

Other possible ways to disperse platinum have recently been explored, using other types of conducting matrix (conducting polymers or ionic membranes). Preliminary work showed the possibility of obtaining significant activity with platinum loadings below 1 mg cm-2, associated with an important decrease in the extent of poisoning phenomena [see also Section II.4(i)].

Similar conclusions were drawn by Aramata and OhniShi for the electrooxidation of methanol on Pt-SPE electrodes. The high activity observed is much more stable over time than that of similar platinized platinum electrodes. By modifying the electrode structure with Sri, Ru, or Ir, it is possible to enhance the activity of Pt-SPE electrodes. Nakajima and Kita studied the direct electrooxidation of methanol from the gas phase on Pt-SPE electrodes. The currents observed were 103 times larger than in the liquid phase, with a smaller extent of deactivation phenomena. When modified with molybdenum atoms, the Pt-SPE electrode exhibits a greater activity, especially at low polarizations (~0.1 V versus RHE).

(f) Analysis of the reaction products

To elucidate the overall reaction mechanism, it is alaso necessary to analyze the products of the reaction. This has generally been performed during prolonged electrolysis using on-line analytical techniques (gas or liquid chromatography) [see Section II.2(i)]. The main reaction product of methanol electrooxidation is carbon dioxide, but formaldehyde and formic acid have also been detected in small amounts. Methyl formate has also been mentioned as a possible by-product.

(g) Conclusions and mechanisms

Since 1980, significant progress has been made in the understanding of the reaction mechanism for the electrooxidation of methanol both in acid and in alkaline medium. The following main points seem to be now widely accepted:

- The decrease of the current densities with time is due to poisoning effects of adsorbed residues coming from the dissociative adsorption of methanol. The species responsible for the electrode poisoning is adsorbed CO, linearly and bridge-bonded to the electrode surface (as proved by *in situ* infrared spectroscopic methods);

- Different adsorbed species exist simultaneously on the electrode surface, and their distribution depends greatly on the surface structure, on the methanol concentration, and on the electrode potential;

- EMIRS studies have allowed the identification of the reactive intermediates as probably CHOads (formyl);

- However, a complete interpretation of the enhancement of the activity of platinum seen with some alloys and some adatoms is difficult, since methanol seems to be a particular case compared to other alcohols or small organics, such as formic acid.

On the basis of all these observations and conclusions, it is now possible to propose a detailed mechanism for the oxidation of methanol at platinum electrodes in acid medium. The steps are as follows:

(1) $$Pt + H_2O \rightarrow Pt-(OH)_{ads} + H^+_{aq} + e^-$$

(2) $$Pt + (CH_3OH)_{sol} \rightarrow Pt-(CH_3OH)_{ads}$$

(3) $$Pt-(CH_3OH)_{ads} \rightarrow Pt-(CH_3OH)_{ads} + H^+_{aq} + e^-$$

(4) $$Pt-(CH_3O)_{ads} \rightarrow Pt-(CH_2O)_{ads} + H^+_{aq} + e^-$$

(5) $$Pt-(CH_2O)_{ads} \rightarrow Pt-(CHO)_{ads} + H^+_{aq} + e^-$$

(6a) $$Pt-(CHO)_{ads} \rightarrow Pt-(CO)_{ads} + H^+_{aq} + e^-$$

(6b) $$Pt-(CHO)_{ads} \rightarrow 2Pt + CO_2 + 2H^+_{aq} + 2e^- + Pt-(OH)_{ads}$$

(6c) $$Pt-(CHO)_{ads} \rightarrow Pt + Pt-(COOH)_{ads} + H^+_{aq} + e^- + Pt-(OH)_{ads}$$

(7a) $$Pt-(CO)_{ads} + Pt-(OH)_{ads} \rightarrow 2Pt + CO_2 + H^+_{aq} + e^-$$

(7b) $$Pt-(CO)_{ads} + Pt-(OH)_{ads} \leftrightarrow Pt + Pt-(COOH)_{ads}$$

(8) $$Pt-(COOH)_{ads} \rightarrow Pt + CO_2 + H^+_{aq} + e^-$$

The key point is the formation of the reactive intermediate $(CHO)_{ads}$ and its further oxidation. There are two main possibilities: $(CHO)_{ads}$ is oxidized to $(CO)_{ads}$ (reaction 6a), leading thus to the poisoning species, or the reactive intermediate $(CHO)_{ads}$ is oxidized either directly to CO_2 (reaction 6b) or through $(COOH)_{ads}$ (reaction 6c), which is further oxidized to CO_2 (reaction 8). The poisoning species $(CO)_{ads}$ can also be oxidized to CO_2, either directly (reaction 7a) or through $(COOH)_{ads}$ (reaction 7b, followed by reaction 8). These different possibilities for the oxidation of $(CHO)_{ads}$ can be summarized in the following scheme,

Charasterization of the Electrode Material

where all the species involved have been observed by EMIRS:

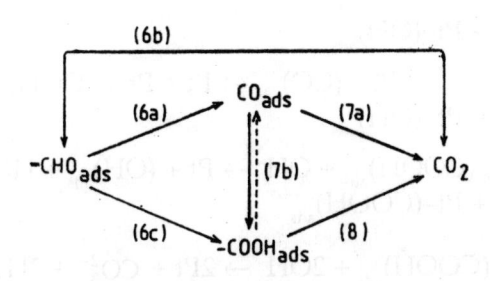

```
                    (6b)
         ┌─────────────────────────┐
         │        CO_ads           │
         │   (6a)  ↑  (7a)         ↓
      -CHO_ads    (7b)            CO_2
         │         ↓               ↑
         │  (6c)  -COOH_ads  (8)   │
         └─────────────────────────┘
```

Figure

From this scheme, it is possible to understand the requirements for the use of methanol in a fuel cell in acid medium. It is necessary to avoid the formation of $(CO)_{ads}$ and to favor route (6b), leading directly to CO_2, or alternatively the indirect routes (6c) and (8) through $(COOH)_{ads}$. This is probably possible by using specific surface structures.

In alkaline medium, the mechanism is similar and could be written as follows:

(1) $\qquad Pt + OH^- \rightarrow Pt-(OH)_{ads} + e^-$

(2) $\qquad Pt + (CH_3OH)_{sol} \rightarrow Pt-(CH_3OH)_{ads}$

(3) $\quad Pt-(CH_3OH)_{ads} + OH^- \rightarrow Pt-(CH_3O)_{ads} + H_2O + e^-$

(4) $\quad Pt-(CH_3O)_{ads} + OH^- \rightarrow Pt-(CH_2O)_{ads} + H_2O + e^-$

(5) $\quad Pt-(CH_2O)_{ads} + OH^- \rightarrow Pt-(CHO)_{ads} + H_2O + e^-$

(6a) $\quad Pt-(CHO)_{ads} + OH^- \rightarrow Pt-(CO)_{ads} + H_2O + e^-$

(6b) $\quad Pt-(CHO)_{ads} + 4OH^- \rightarrow Pt-CO_3^{2-} + 3H_2O + 2e^-$
$\qquad\qquad + Pt-(OH)_{ads}$

(6c) $\quad Pt-(CHO)_{ads} + OH^- \rightarrow Pt + Pt-(COOH)_{ads}$
$\quad\quad\quad + H_2O + e^- + Pt-(OH)_{ads}$

(7a) $\quad Pt-(CO)_{ads} + 3OH^- \rightarrow 2Pt + CO_3^{2-} + 2H_2O + e^-$
$\quad\quad\quad + Pt-(OH)_{ads}$

(7b) $\quad\quad Pt-(CO)_{ads} \leftrightarrow Pt + Pt-(COOH)_{ads}$
$\quad\quad\quad + Pt-(OH)_{ads}$

(8a) $\quad Pt-(COOH)_{ads} + OH^- \rightarrow Pt + (OH)_{ads} + HCOO^-$
$\quad\quad\quad + Pt-(COOH)_{ads}$

(8b) $\quad Pt-(COOH)_{ads} + 2OH^- \rightarrow 2Pt + CO_3^{2-} + 2H_2O$
$\quad\quad\quad + Pt-(OH)_{ads}$

The main difference with respect to the mechanism in acid medium is the formation of formate and carbonate ions (steps 6b, 7a, 8a, and 8b). The amount of $(CO)_{ads}$ formed is smaller than in acid medium, leading to a greater electrocatalytic activity.

(ii) Formaldehyde

The electrooxidation of formaldehyde has been often studied as a model reaction. Only four electrons are required for complete oxidation to carbon dioxide. Formaldehyde is also an intermediate oxidation product found during the electrocatalytic oxidation of methanol.

If the reaction mechanism seems, *a priori,* rather simple with only a few possible adsorbed intermediates, there are difficulties in studying the electrooxidation of formaldehyde due to the lack of stability of this molecule. Formaldehyde is generally supplied in aqueous solution, stabilized by methanol. Obviously, the presence of methanol may present some problems for the determination of the true reactivity of the aldehyde during its oxidation, even if it is much more reactive than methanol. Formaldehyde is often available as paraformaldehyde, that is, its polymeric form. The only way to obtain the pure monomer from the polymeric form is to reflux for several hours and to rapidly use the monomer obtained.

One of the first complete studies on the electrooxidation of formaldehyde, in comparison with that of formic acid and methanol, was done by Buck and Griffith '54 whereas Sibille et al. and Van Effen and Evans compared the behavior of several aldehydes. Several other recent works can be mentioned, such as those of Beltowska-Brzezinska and Heitbaum on gold-platinum alloys in alkaline medium and those of Avramov-Ivic et al. on noble metals. Other kinds of electrode material have been considered, such as copper, copper-nickel alloys, and even ternary alloys, Cu/Pd/Zr. The effect of adatoms was also considered, and possible structural effects were investigated by using platinum and gold single-crystal electrodes.

(a) Oxidation on pure metals

Important differences in the electroreactivity of formaldehyde exist depending on the nature of the electrode material and on the pH of the electrolyte. On platinum in acidic medium, the voltammogram of formaldehyde electrooxidation is characterized by a significant hysteresis between the positive and the negative sweeps (Fig. 30a). This fact is due to extensive poisoning of the electrode surface, and again the species responsible for the poisoning has been clearly identified by in situ infrared spectroscopy as adsorbed CO.

A mechanism similar to that given for methanol oxidation can be proposed for the electrooxidation of formaldehyde in acid medium:

(1) $$Pt + H_2O \rightarrow Pt-(OH)_{ads} + H^+_{aq} + e^-$$

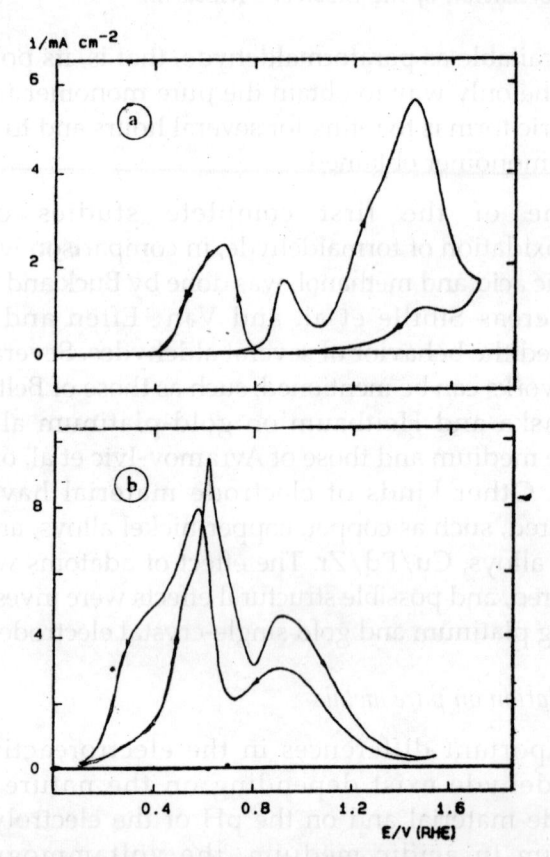

Figure 12. Voltammograms showing the oxidation of 0.1 M formaldehyde at a platinum electrode (25°C, 50mVs⁻¹) in: (a) 0.1 M HClO$_4$; (b) 1 M NaOH.

(2) $\quad$ Pt + (HCHO)$_{sol}$ → Pt − (HCHO)$_{ads}$

(3) $\quad$ Pt − (HCHO)$_{ads}$ → Pt − (CHO)$_{ads}$ + H$^+_{aq}$ + e⁻

with step (4) and subsequent steps being similar to steps (6) to (8) for methanol oxidation in acid medium.

This mechanism is derived from that proposed earlier by Spasojevic *et al.*, but differs in the nature of the poisoning species, which is here (CO)$_{ads}$, whose formation originates in the same process as in the case of methanol electrooxidation.

In alkaline medium, the electrooxidation of HCHO occurs at a significant rate on all noble metals (Au, Pd, Pt, Rh, but also Ir and Ag), with no serious poisoning effect (see Fig. 30b for platinum). Gold appears to be the most active electrocatalyst.

At alkaline pH, HCHO is known to exist as a gem-diol form, which results from the following reaction:

$$HCHO + OH^- \leftrightarrow H_2C(OH)O^-$$

This form is probably the main form adsorbed on the metal electrode, so that, under these conditions, formate is the main oxidation product.

However, for use in an electrochemical energy conversion device, alkaline medium is not interesting, because of the consumption of electrolyte by salt formation.

Mechanisms of oxidation of formaldehyde in alkaline medium have been described by Beltowska-Brzezinska and Heitbaum. The overall reaction, either on Au or Pt electrodes, can be written as

$$HCHO + 3OH^- \rightarrow HCOO^- + 2H_2O + 2e^-$$

or, if the *gem*-diol form is considered, as

$$H_2C(OH)O^- + 2OH^- \rightarrow HCOO^- + 2H_2O + 2e^-$$

Formate ions are known not to be electroreactive in alkaline medium, so that the formation of carbon dioxide (in fact, CO_3^{2-}) is negligible.

The detailed mechanism proposed by Beltowska-Brzezinska and Heitbaum takes into account the possible formation of different adsorbed intermediates, including CO. The formation of such a poison was demonstrated by

Avramov-Ivic et al. in alkaline medium and by Nishimura et al. in acid medium.

The influence of the structure of the electrode surface was investigated by Adzic et al. The differences in electrocatalytic activity observed between the different orientations of platinum (or gold) single crystals remain small in comparison with those observed in acid medium for other molecules. This general behavior is similar to that of methanol on Pt(hkl) electrodes in alkaline medium.

(b) Effect of adatoms

The poisoning phenomena observed, mainly in acid medium, during the electrooxidation of formaldehyde can be drastically reduced by using adatoms to modify the electrode surface.

In acid medium, the oxidation of HCHO on platinum is considerably enhanced by underpotential deposition of Pb, Bi, or Ti. According to Motoo and Shibata, two kinds of adatoms can be distinguished: the first type of adatoms (Cu, Ag, Tl, Hg, Pb, As, Bi) have only a geometrical effect, whereas the second kind of adatoms (Ge, Sn, Sb), which are able to adsorb oxygen at low potentials, give rise to greater enhancement factors. The authors interpreted the enhancement effects in terms of three reaction paths:

$$HCOOH \xrightarrow{-2} \text{Poisoning species}$$
species

$$HCHO \longrightarrow HCHO \xrightarrow{\quad 3 \quad} CO_2$$
in solution   species

with path 1 above.

If the number of neighboring platinum sites is greater than three, the formation of poisoning species is favored. If their number is equal to three, which can be obtained by underpotential deposition of some adatoms (e.g., Bi), the formation of poisons is inhibited, and path I is favored. With adatoms that adsorb oxygen, the direct path 3 seems to be favored.

(iii) Formic Acid

As mentioned above, the possible use of methanol in fuel cells explains why the electrooxidation of methanol has been the most studied reaction involving a one-carbon organic molecule. However, formic acid, which is also an intermediate product of methanol oxidation, has also been considered. Thus, with the aim of elucidating the mechanism of the oxidation of methanol, various research groups have carried out parallel studies on the electrooxidation of formic acid.

Formic acid is oxidized in acid medium according to the following overall reaction:

$$HCOOH \rightarrow CO_2 + 2H^+ + 2e^-$$

Platinum is a good electrocatalyst for the oxidation of formic acid, but rhodium and especially palladium also have good activity in acid medium. The reactivity of formic acid (or formate) varis dramatically with the pH of the solution. In neutral medium, at a pH corresponding to the pK_a of formic acid ($pK_a \approx 4$), gold exhibits significant activity, although it is practically completely inactive in acid medium. Moreover, platinum, palladium, and rhodium also display increased activity in neutral medium.

In acid medium, platinum remains the most studied electrocatalyst for formic acid oxidation.

(a) Nature of the adsorbed species

Parsons and VanderNoot[12] reviewed the electrooxidation of, HCOOH on Pt and proposed a general scheme:

$$\text{HCOOH} \begin{array}{c} \text{Reactive intermediate} \to CO_2 \\ \\ \text{Poisoning species} \to CO_2 \end{array}$$

Capon and Parsons postulated –COOH as the reactive adsorbed intermediate and –COH as the poisoning species. These conclusions were drawn from pure electrochemical measurements, aimed mainly at determining the quantity of electricity associated with the oxidation of the adsorbed species.

Similarly, Kazarinov et al., on the basis of radiometric measurements, also proposed the presence of adsorbed COH as a poisoning species. More recently, Beden et al., using in situ infrared reflectance spectroscopy (EMIRS), proved unambiguously the occurrence of adsorbed CO on Pt after adsorption of formic acid. Two IR absorption bands attributed to adsorbed CO were detected, one assigned to linearly bonded CO and the other to bridge-bonded CO (Fig. 16c). In order to confirm these assignments, the EMIRS spectrum of the adsorbed species resulting from the adsorption of gaseous CO, dissolved in the electrolytic solution, were recorded under the same experimental conditions (Fig. 16h). These experiments showed that the same poisoning species were involved during the adsorption of CO and that of both formic acid and methanol (Fig. 16).

On the other hand, Heitbaum and co-workers, using DEMS and XPS measurements, concluded that another inhibiting species, namely, $(HCO)_{ads}$, was present on the electrode surface during adsorption of HCOOH. They explained the presence of $(CO)_{ads}$, as detected by EMIRS, as the consequence of the reaction of their postulated (HCO)ads with residual oxygen in the electrochemical cell.

Corrigan and Weaver also observed (CO)ads in performing infrared measurements by SPAIRS [see Section II.3(iv.e)]. They concluded from their experiments that $(CO)_{ads}$ is the species responsible for the poisoning phenomena.

(b) Structural effects

The occurrence of different kinds of adsorbed species during the oxidation of formic acid was also indirectly proved by studies carried out on electrodes with different surface structures. The extensive literature on this subject may be classified into three categories according to the type of electrode used: single crystals, alloys, and platinum electrodes modified by foreign adatoms.

Single crystals. The best way to observe structural effects in electrocatalysis is to use model electrode surfaces, such as single crystals. After the first experiments, in 1976, on gold single crystals, various research groups observed important structural effects during the electrooxiclation of formic acid on platinum single crystals or on platinum single crystals modified by adatoms. Even though the findings reported by these research groups for platinum single crystals are not all in agreement, the influence of the surface structure during the electrocatalytic oxidation of formic acid was clearly demonstrated (Fig. 31).

The current densities for formic acid oxidation show a dependence on the orientation of the Pt(*hkl*) planes, and the decrease in the current that results from electrode poisoning depends strongly on the nature of the face studied. This poisoning effect is drastic for the Pt(110) plane, with a rapid decrease of the current versus time. Conversely, it is much smaller for Pt(111), which appears to be less sensitive to poisoning species, even though the initial activity is rather low. The Pt(100) plane is blocked until the potential reaches about 0.7 V versus RHE, the

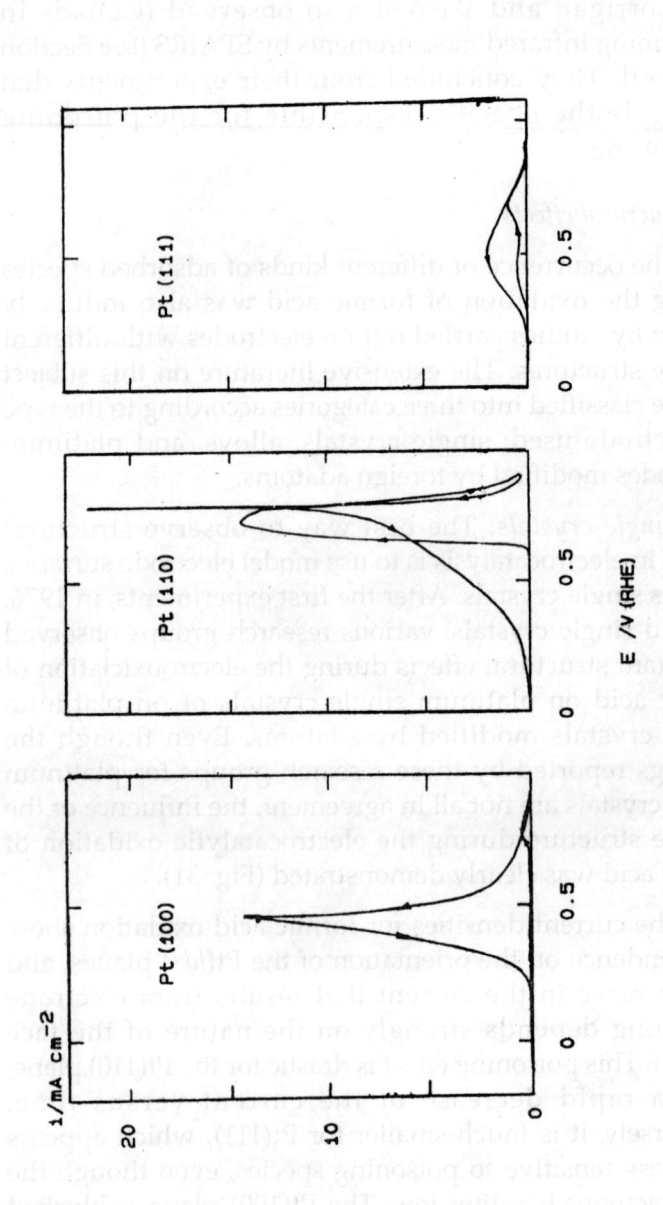

Figure 13. Voltammograms of low-index platinum single crystals showing pronounced structural effects in the electrooxidation of formic acid (0.1 M HClO$_4$, 0.1 M HCOOH, 25°C, 50 mBs^{-1}, first sweep).

current attaining appreciable values only during the negative sweep. This behavior is a little bit different from that of methanol on P000) [see Section III.1 (i.c)]

In preliminary studies, the influence of adatoms on single-crystal electrodes was investigated in the case of platinum. As expected, important effects were observed with lead and bismuth adatoms.

Alloys. Historically, the first attempts to modify the structure of a catalytic electrode involved the use of metallic alloys. The electrooxidation of formic acid has been investigated on different types of alloys: Pt/Au, Pt/Pd, Pt/Rh, and Au/Pd. Except for Pt/Au, these alloys are solid solutions, and it is possible to vary continuously the bulk composition of the electrode. One of the main problem remains the determination, with good accuracy, of the surface composition, which differs significantly from the bulk composition, due to surface enrichment by preferential dissolution of one metal component. Techniques to estimate the true surface composition are discussed in Section II.4(ii.c). Synergistic effects were observed in many cases: the electrode activity for some surface compositions is greater than that for the pure metals. Such effects were interpreted by taking into account the nature of the adsorbed species: strongly bonded species (poisons), for instance, are more easily formed on platinum than on gold or palladium electrodes. By modification of the platinum electrode with a second metal, the platinum sites are diluted and the formation of such poisons is less favored.

Adatoms. The electrooxidation of HCOOH is probably the most typical example of electrocatalytic reactions used to illustrate the influence of adatoms on the activity of platinum. Adzic has reviewed this subject.

Striking catalytic effects are observed in the case of the oxidation of HCOOH on Pt, Rh, Ir, and Pd in the presence of adatoms, such as Pb, Bi, Tl, and Cd. A very significant enhancement effect on HCOOH oxidation at platinum electrodes is observed in the presence of lead adatoms. During the positive sweep in the range 0.2 to 0.8 V versus RHE, only a small peak is observed on pure platinum. In the presence of small amounts of lead salts in solution, the oxidation of HCOOH during the positive sweep leads to a large peak (of around 70 mA cm^{-2}). Another important observation is the quasi-superposition of the oxidation currents during the forward and the reverse sweeps in the

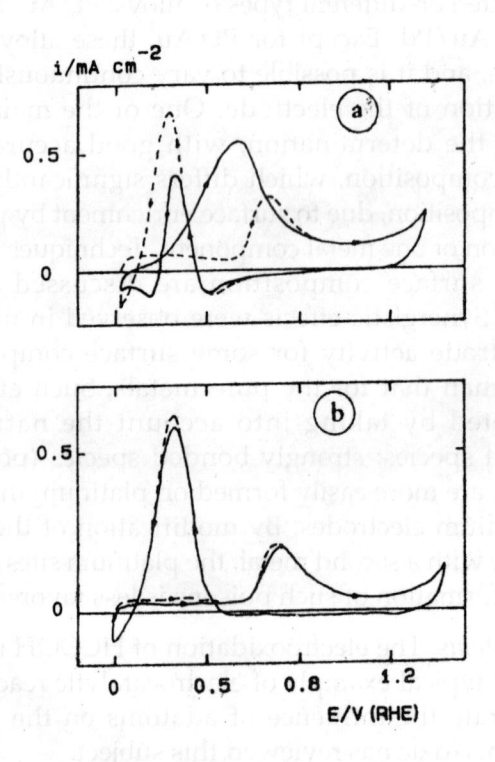

Figure 14. Effect of lead and cadmium adatoms on the voltammograms showing the oxidation of 0.1 M HCOOH in 0.5 M HCO4 at Rh electrodes (250C, 50mVs-1). The dashed curves correspond to unmodified Rh electrodes. (a) 5 × 10^{-4} M Pb^{2+}; (b) 10^{-3} M Cd^{2+}.

range 0.2-0.5 V versus RHE, a fact which confirms the dramatic decrease in the poisoning phenomena. This spectacular effect is also observed with systems such as Pt/Cd$_{ads}$, Pt/Tl$_{ads}$, and Pt/Bi$_{ads}$, but to varying extents. However, the effect of cadmium is somewhat different. The maximum current densities observed are not affected as much relative to those on pure platinum, but the potential at which oxidation occurs is shifted negatively.

The effect of these adatoms is also illustrated in the case of formic acid oxidation at rhodium electrodes, modified by lead or cadmium adatoms (Fig. 32). As with platinum, lead adatoms give enhancement effects, but, conversely, cadmium adatoms do not change the electrocatalytic activity of rhodium. An explanation of this different behavior has been given on the basis of IR spectroscopic measurements (see below).

(c) *Discussion of the reaction mechanism: Example of the rhodium electrode*

The electrooxidation of formic acid can be assumed to follow a dual-path mechanism:

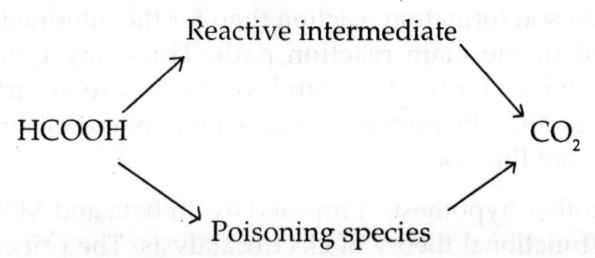

$$\text{HCOOH} \begin{matrix} \nearrow \text{Reactive intermediate} \searrow \\ \searrow \text{Poisoning species} \nearrow \end{matrix} \text{CO}_2$$

Hence, the reactive intermediate is of the (COOH)$_{ads}$ type, and, as stated above, the poisoning species is now known to be adsorbed CO. Different conclusions about the exact nature of the poisoning species led to different hypotheses being put forward to explain the effect of adatoms. According to Bagotsky and Vassiliev and, later,

Adzic, the structure of the poisoning species is $(COH)_{ads'}$ which thus needs Hads to be formed. However, no clear experimental confirmation of the occurrence of this poisoning species was obtained at that time. According to the original explanation by Adzic, the foreign adatoms block the surface active sites and prevent the formation of adsorbed hydrogen. In the absence of $H_{ads'}$ according to Adzic, the formation of the postulated poisoning species, $(COH)_{ads'}$ would be impossible. This hypothesis fails as the poisoning species is now known and well accepted to b $(CO)_{ads'}$ and as this latter species obviously does not need Had, to be formed. Another remark is worth making: a platinum electrode can be poisoned even when it has been maintained at potentials more positive than those required for the adsorption of hydrogen

Various other hypotheses were proposed by several authors. Hartung et al. and Kokkinidis suggested that adatoms work by blocking the poison formation reaction. According, to this hypothesis, a certain number of sites of a give symmetry are required to allow the formiation of a poisoning species. Generally, this number of sites is greater for the poison formation reaction than for the intermediate involved in the main reaction path. Thus, any type of adatoms which form small islands on the electrode surface should enhance the electrocatalytic activity of the substrate, which is not the case.

Another hypothesis, proposed by Shibata and Motoo, is the bifunctional theory of electrocatalysis. The principle is rather simple. A catalytic surface partially covered with foreign adatoms presents two types of active sites-firstly, the sites of the catalytic metal, namely, platinum, which are able to adsorb and dissociate the organic molecules, and, secondly, the adatom sites, which can adsorb the oxygen atoms or the hydroxyl ions necessary for adsorbed CO to be oxidized to carbon dioxide. If the potential at which oxygen species are adsorbed is lower for the adatoms than for the platinum sites, enhancement of the reaction rate at lower potentials becomes possible.

Finally, adatoms may also act by altering the electronic properties of the substrate or by behaving as redox intermediates. This explanation is more satisfying for alkaline solutions, where no significant poisoning effects are observed, nor can a $(CO)_{ads}$ contribution be invoked.

It is really difficult to definitively choose one of these different hypotheses. However, the following experimental facts about the electrooxidation of formic acid at noble metals are now widely accepted: (i) adsorbed CO (linearly and bridge bonded) is the poisoning species; and (ii) there is no experimental evidence for the presence of adsorbed COH species. Only some adatoms have positive effects on the electrooxidation of formic acid; among these, lead and bismuth have the most significant effects. They presumably may occupy some of the specific electrode active sites necessary for formation of the poisoning species, but at the same time they may behave according to a bifunctional mechanism as well. Modifications of the electronic properties cannot be excluded either, at least in some cases. There is, to date, no definitive experimental proof that favors one hypothesis or the other, except that the modification of the electrode coverage by strongly bonded species is firmly demonstrated. Recent results concerning the Rh/ HCOOH system with lead, bismuth, or cadmium adatoms, obtained by in situ IR spectroscopic measurements (EMIRS), gave interesting information. The modification of the surface distribution of adsorbed species during adsorption of an organic molecule at adatom-modified electrodes was detected for the first time. Thus, the degree of coverage by bridge-bonded CO species was found to be much more affected by lead adatoms than was that by linearly bonded CO species (Fig. 33). It was concluded that the poisoning effect was due not only to the presence of both CO species (linearly and bridge-bonded), but above all to interactions between these two kinds of species. The inhibition of the formation of bridge-bonded CO is sufficient to enhance the catalytic activity of the

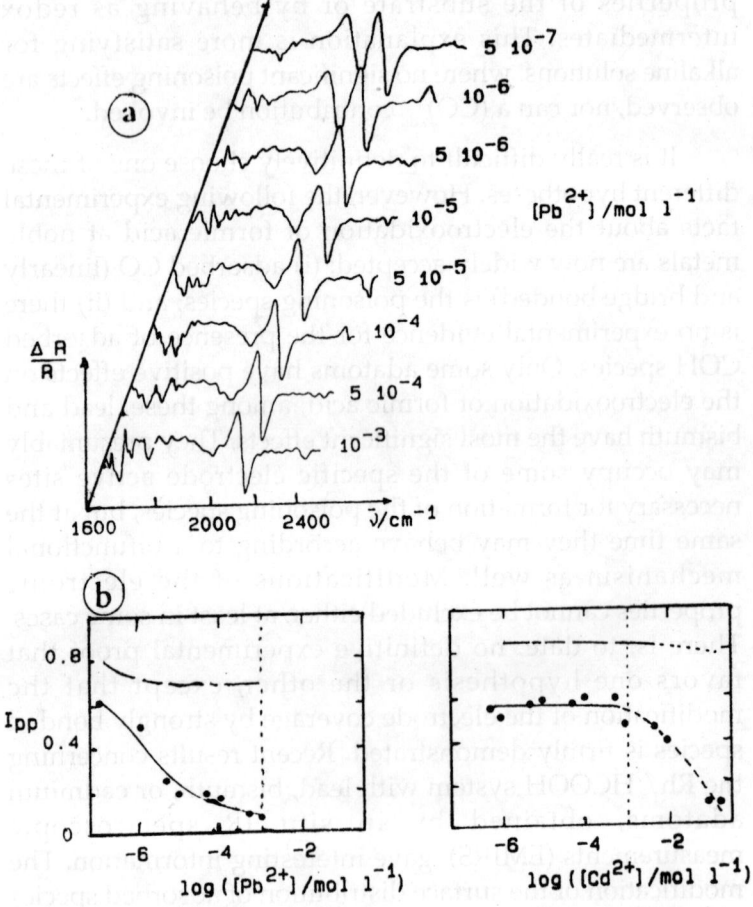

Figure 15. (a) Effect of lead adatoms on the EMIRS spectra of Co$_{ads}$ resulting from the adsorption of 0.1 M HCOOH in 0.5 M HClO$_4$ at rhodium electrodes. (b) Plots of the peak-to-peak intensities of the EMIRS bands of Coads resulting from HCOOH adsorption at modified rhodium electrodes as a function of the bulk concentration of the precursor salt (Pb^{2+} or Cd^{2+}).

electrodes. The same observation was made with bismuth-modified rhodium electrodes. On the other hand, the case of cadmium is completely different, in that cadmium adatoms have nearly no effect on HCOOH oxidation at rhodium electrodes. An EMIRS study showed no modification of the distribution of the adsorbed CO species in the presence of Cd adatoms on the electrode surface. It seems that Cd adatoms are not adsorbed strongly enough to displace the adsorbed CO species (Fig. 33b).

All the results now accumulated, particularly those obtained with well-defined electrodes and those obtained using in situ spectroscopic methods (e.g., EMIRS), allow a detailed mechanism to be proposed for the electrooxidation of formic acid at noble metals. This can be illustrated by the case of a rhodium electrode in acid medium, as a typical example.

The main reaction path is:

(1) $$Rh + H_2O \rightarrow Rh-(OH)_{ads} + H^+_{aq} + e^-$$

(2) $$Rh + (HCOOH)_{sol} \rightarrow Rh-(HCOOH)_{ads}$$

(3) $$Rh-(HCOOH)_{ads} \rightarrow Rh-(COOH)_{ads} + H^+_{aq} + e^-$$

(4) $$Rh-(COOH)_{ads} \rightarrow Rh + CO_2 + H^+_{aq} + e^-$$

The poisoning species, linearly and bridge-bonded CO, are produced by the following reactions:

(5a) $$2Rh + HCOOH \rightarrow \begin{matrix}Rh\\ \diagdown\\ \diagup\\ Rh\end{matrix}(CO)_{ads} + H_2O$$

(5b) $$Rh + Rh-(COOH)_{ads} \leftrightarrow Rh-(CO)_{ads} + Rh-(OH)_{ads}$$

(6) $\text{Rh}_2\text{(CO)}_{ads} \leftrightarrow \text{Rh-(CO)}_{ads} + \text{Rh}$

In the presence of lead adatoms, the formation of bridge-bonded CO is inhibited, which leads to an enhancement of electrocatalytic activity.

The oxidation of the poisoning species can be summarized by the following reactions:

(7a) $\text{Rh-(CO)}_{ads} + \text{Rh-(OH)}_{ads} \rightarrow 2\text{Rh} + CO_2 + H^+_{aq} + e^-$

(7b) $\text{Rh}_2\text{(CO)}_{ads} + \text{Rh-(OH)}_{ads} \rightarrow 3\text{Rh} + CO_2 + H^+_{aq} + e^-$

(iv) Carbon Monoxide and Reduced CO_2

Although, strictly speaking, CO is not an organic molecule, it is worth considering here, due to its virtual omnipresence as an intermediate or a product of most adsorption . and electrooxidation processes involving small oxygenated organic molecules. Exceptions to this may arise in the case of long-chain aliphatic acids, as demonstrated by Horanyi,[91] using radiotracer techniques, and, more recently, by Leung and Weaver on the basis of an infrared and Raman spectroscopic survey. In most cases, as for the exidation of methanol [see Section III.1 (I)] adsorbed CO acts as a poison on catalytic surfaces, drastically decreasing their activity. This conclusion is now widely accepted, even though some research groups have postulated, from pure electrochemical measurements, that differences exist between species arising from the adsorption of gaseous CO and species arising from. the dissociative adsorption of small oxygenated organic molecules. On the other hand, the ease with which CO is formed and adsorbed on almost all catalytic surfaces led Beden et al. to emphasize recently the role that CO may play as a probe molecule to test the electrocatalytic activity of metallic electrodes.

Since the work of Breiter, it has been known that the electrooxidation of CO on polycrystalline noble metals is a complex reaction, leading to multiple voltammetric peaks in a potential range depending to a large extent on the adsorption potential and on the concentration of dissolved CO in the electrolytic solution. Usually, with a solution saturated in dissolved CO, and for adsorption potentials greater than 0.4 V versus RHE (but below 0.6-0.7 V), only one narrow oxidation peak is detected at 0.9 V. If the adsorption potential is chosen below 0.4 V, multiple peaks are obtained. Two main peaks are usually observed: the more cathodic one at around 0.45 V, and the second one at around 0.8-0.9 V. More recently, following a study by Kita and Nakajima, it has been recognized by Caram and Gutierrez that the behavior of CO depends strongly on its admission potential, that is, on the potential at which it has been introduced into the previously deareated cell (the value of 0.22 V versus RHE being especially critical).

The formation of an adsorbed CO layer results from the contributions of at least three distinguishable types of adsorbates, of different configurations, on the surface, depending on the crystallographic sites. Thus, linearly bonded CO_L is adsorbed "on top" of sites, while bridge-bonded CO_B lies between two sites, and multi-bonded CO_m occurs at higher coordination sites [see Section II.3(iv.c)]. However, depending on potential and/or on coverage, other configurations may have to be considered as well. For instance, Ikezawa et al. reconsidering the infrared band intensities, postulated that part of the surface might be covered by CO species adsorbed in a flat position on two adjacent Pt atoms. Such species would be IR inactive, which would explain their nondetection.

As stated above, the electrooxidation process of CO is complex. On Pt, or on Rh, it is known that the electrochemical behavior of CO depends on whether it was previously adsorbed in the double-layer region or in the hydrogen region. Kunimatsu *et al.*, for instance, have shown some interesting differences in the CO infrared bands when adsorption on polycrystalline Pt was carried out at two separate potentials, one close to the hydrogen evolution potential, and the other in the middle of the double-layer region. This suggests that some of the CO species may react with hydrogen.

Recent work on Pt single crystals demonstrated the sensitivity of CO adsorption to surface structure. Moreover, extensive spectroscopic investigations of CO adsorption on well-defined single crystals of Pt or Rh by Weaver and coworkers revealed its dependence on surface reconstruction, through a comparison of (hkl) ordered and disordered crystal faces.

Of interest for practical applications of electrocatalysis is the fact that surface roughness acts to diminish the adsorption of CO, particularly when it results from chemisorption. This was known from fuel cell studies: developed electrodes are less poisoned than model smooth electrodes. It has also been recently demonstrated using EMIRS with preferentially oriented Pt electrodes, as well as with electrochemically developed Pt surfaces, obtained by using high-frequency pulse potential programs.

Direct oxidation of gaseous CO can be realized at metallized electrodes whose opposite side is in contact with an electrolytic solution. In an earlier work, Gibbs *et al.* used a PTFE membrane onto which a gold or a platinum layer was sputtered. The electrochemical behavior of these electrodes was found to be similar to that of conventional electrodes. In a more recent work, Kita and Nakajima obtained much higher currents at Au-SPE electrodes than at smooth gold, while less poisoning was observed.

In a similar way to the oxidation processes involving small organic molecules, the electrochemical oxidation of CO may be enhanced either by using adatoms or by using alloys of noble metals. Thus, in a long series of papers, Motoo and co-workers have examined many possibilities involving Pt, Rh, Au, and Ir electrodes, and their modification by Ru, As, and Sn adatoms (see, for instance, Refs. 472-476).

As mentioned above, adsorbed CO is the species responsible for electrode poisoning during the electrooxidation of numerous small oxygenated organic molecules. However, the case of the electrooxidation of the adsorbed species involved in the electroreduction of carbon dioxide has led to some controversy concerning the exact nature of the species formed in the hydrogen adsorption region. According to various authors, these species are somewhat different from those formed from dissolved gaseous CO, and a clathrate-type structure has been proposed by Marcos *et al.* However, confirming the early suggestion of Breiter, made on the basis of pure electrochemical measurements, Beden *et al.* found the same IR absorption bands for reduced CO_2 species and for adsorbed CO coming from dissolved CO, CH_3OH, HCOOH, and HCHO. The confirms that adsorbed CO is probably also formed during the electroreduction of carbon dioxide.

(v) Summary of the Reaction Mechanism

To summarize the preceding discussion about the adsorbates formed from C_1 organic compounds on noble metals, Fig. 34 illustrates the more likely pathways between

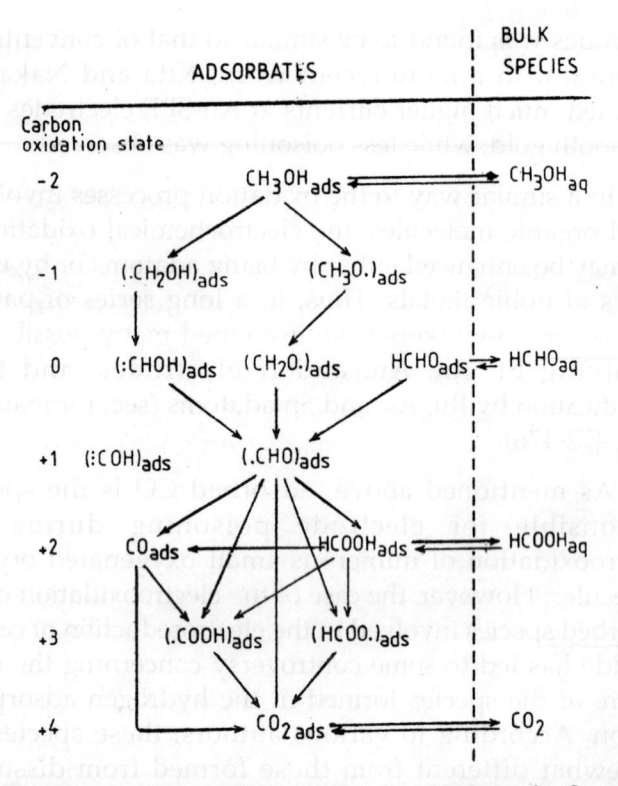

```
                              |  BULK
         ADSORBATES           |  SPECIES
_____
Carbon
oxidation state
   -2           CH₃OH_ads  ⇌  CH₃OH_aq

   -1    (.CH₂OH)_ads   (CH₃O.)_ads

    0    (:CHOH)_ads  (CH₂O.)_ads   HCHO_ads ⇌ HCHO_aq

   +1    (:COH)_ads    (.CHO)_ads

   +2    CO_ads        HCOOH_ads  ⇌  HCOOH_aq

   +3    (.COOH)_ads   (HCOO.)_ads

   +4              CO₂ ads  ⇌  CO₂
```

Figure 16. Pathways between bulk species and adsorbates for C_1 organic compounds with different oxidation state of carbon.

bulk species and adsorbates, from methanol to CO_2. In the redox reference scale used the carbon in methane (which is not represented, here) would have an oxidation number of -4, while the carbon in carbon dioxide has an oxidation number of $+4$.

Oxidation of Monofunctional Molecules

Numerous oxygenated molecules with more than one carbon atom have been considered both as fuels in fuel cells and its raw materials for electrosynthesis. In this section, the oxidation of such molecules will be discussed, with selected examples concerning C_2 to C_6 molecules. Different aspects will be considered—for example, the nature of the functional group (alcohol, carboxylic acid, aldehyde, or ketone), the length of the carbon chain, and he type of isomer.

A preliminary remark concerns the possible use of such molecules in fuel cells. An increase in the number of carbon atoms leads to an increase in the number of possible intermediates and reaction products in the course of their electrooxidation, and consequently, to more complex reaction mechanisms. Furthermore, they are much more difficult to oxidize completely to carbon dioxide, so that their practical energy density is much lower than expected (e.g., complete oxidation of glucose would liberate 4.4 kW-h kg^{-1}, whereas partial oxidation to gluconic acid gives only $\frac{1}{12}$ of this energy, i.e. 0.4 kW-h kg^{-1}). Finally, the prices of these fuels (e.g., ethanol, ethylene glycol) are much higher than that of methanol, and this may limit their practical use.

However, from a fundamental point of view, studies of the electrocatalytic oxidation of such products are quite useful for understanding the general reaction mechanisms. Moreover, mainly for multifunctional molecules, their application in the field of electrosynthesis is a further motivation for such studies.

(i) Reactivity of Alcohol Functions

Due to the good solubility of alcohols in water, their electrooxidation has been widely studied in aqueous electrolytes. Two main topics have been investigated: the effect of the length of the carbon chain, and the influence of the position of the functional group in the carbon skeleton.

Beginning with the primary alcohols (characterized by the -CH_2OH functional group), the electrocatalytic oxidation of the first member of the series, ethanol, has been the subject of numerous studies, but that of propanol and of butanol have also been considered. Several reviews of work on the electrooxidation of several alcohols on platinum and platinum-gold alloys have appeared. Gold was also considered as a good electrode catalyst.

Submonolayers of various adatoms were also considered to modify the electrocatalytic behavior of platinum in the oxidation of several alcohols

From all these studies, although performed under different experimental conditions, making comparison of the results difficult, some general conclusions can be drawn:

• The rate of alcohol electrooxidation depends strongly on the pH of the solution. Platinum is the best electrocatalyst in acid medium, whereas gold is quite inactive. Conversely, in alkaline medium, gold is usually a very active catalyst for the electrooxidation of alcohols at rather high potentials.

• The reactivity of alcohols depends on the position of the functional group. On platinum, primary alcohols are more reactive than secondary alcohols and tertiary alcohols are practically unreactive at room temperature. However, in this latter case, a small reactivity is observed at higher temperatures in a potential region corresponding to the adsorption of oxygen. This reactivity, even though small, is very interesting and unexpected. For primary and secondary alcohols, the removal of a hydrogen in the a position is the first step in dissociative adsorption.[109] Since tertiary alcohols do not have any H atom bonded to the carbon in the a position, the mechanisms of their electrooxidation must be completely different from those for primary or secondary alcohols.

• Gold is a very active catalyst for the electrooxidation of alcohols in alkaline medium. The reactivity of primary

Charasterization of the Electrode Material

alcohols is usually lower than that of secondary alcohols. This difference is probably related to the inductive effect of the adjacent methyl groups present in secondary alcohols. Conversely to what happens at platinum electrodes, the reactivity of alcohols at gold electrodes increases with the number of carbon atoms, at least in a series up to six carbons.

- Values of the apparent activation energies are rather similar for the electrooxidation of all alcohols between one and four carbons in length and are in line with those expected for adsorption processes.

- The first step in the electrooxidation on platinum is an adsorption process with, in most cases, dissociation of the molecule, mainly at low potentials in the double-layer region. This dissociative process is obvious on platinum, but does not seem to occur on gold electrodes.

- As for the oxidation of methanol, poisoning effects are observed for all alcohols during their electrooxidation on platinum. The nature of the adsorbed residue produced during the dissociative chemisorption of alcohol molecules has been studied by in situ infrared reflectance spectroscopy. According to experiments using EMIRS and SNIFTIRS, the nature of the poisoning species is definitively $(CO)_{ads}$ formed during the dissociation of the alcohol molecule. This is true for all alcohols and even for aldehyde compounds. Differences exist only in the extent of coverage of the surface by adsorbed CO, which is higher for the smallest molecules, reaching 0.9 for methanol.

The general reaction mechanism of the oxidation of aliphatic alcohols on platinum can be summarized with the following simple scheme:

Alcohol
$\quad$ Poisoning species $(CO_{ads}) \to CO_2$

$\quad$ Reactive intermediates $\to$ Products (aldehydes, ketones, carboxylic acids)

The nature of the reactive intermediates is more difficult to identify clearly. In the case of ethanol, an aldehyde-like intermediate ($CH_3-\underset{C}{}=O$) is postulated.

In general, the main product formed during oxidation of primary and secondary alcohols is the corresponding aldehyde and ketone, respectively:

$$R-CH_2OH \to R-CHO + 2H^+_{aq} + 2e^-$$

$$R-CHOH-R' \to R-CO-R' + 2H^+_{aq} + 2e^-$$

Further oxidation of aldehydes is always possible, lead usually to carboxylic acids:

$$R-CHO + H_2O \to R-COOH + 2H^+_{aq} + 2e^-$$

This will depend on the potential applied to the electrode. The structure of the electrode support may also influence the nature of the products formed.

As a typical example, the electrocatalytic oxidation of ethanol at a platinum electrode gives mainly acetaldehyde and acetic acid, as proved by chromatographic analysis. A very small amount of CO_2 is detected in the gas phase, coming probably from the oxidation of the CO poisoning species.

Role of the electrode material

$\quad$ *Alloys.* As mentioned before, the electrooxidation

mechanisms on gold and platinum seem to be different: no poisoning effect is observed on gold, but oxidation occurs at higher potentials than on Pt. The use of platinum-gold alloys thus could offer the possibility of avoiding electrode poisoning, but with a better catalytic activity than exhibited by platinum. However, in acid medium, no synergistic effect was observed, and the activity of Pt is only decreased by the addition of gold. In alkaline medium, significant synergistic effects are observed with the alloy 20: 80 Pt/Au.

Adatoms. Small, but significant, effects were observed during the oxidation of alcohols at adatom-modified electrodes The oxidation potential is generally lowered, and, in the most favorable cases, namely, in the presence of adatoms of lead, germanium, thallium, or bismuth, the maximum activity is increased by a factor of 2 to 3. A more detailed study of ethanol oxidation showed that the two steps (oxidation to acetaldehyde and then to acetic acid) should be considered separately. The first step is inhibited by the presence of some adatoms (Ag, Hg, Se, Te, and Bi), but the second one is activated by some of these same adatoms (Se, Te, and Bi).

Single crystals. In the electrooxidation of molecules with one carbon atom, important structural effects were demonstrated (see Section III.1). Only a very few studies have been carried out to investigate such effects during the oxidation of alcohols heavier than methanol. The electrooxidation of ethanol on Pt(hkl) single crystals shows structural effects similar to those encountered in the case of methanol in acid medium, but with smaller currents. Moreover, in alkaline medium, the three low-index planes of platinum appear to be much more active than polycrystalline platinum, conversely to the case of methanol. The same observations were also made in the case of the oxidation of butanol at Pt(*hkl*) electrodes in alkaline medium. Thus, it has not been possible to simulate the electrocatalytic behavior of polycrystalline platinum by that of the three low-index single crystal planes.

(ii) Reactivity of the Aldehyde Function

The electrocatalytic oxidation of aldehydes has been much less studied than that of alcohols, except that of formaldehyde [see Section III.1(ii)] Nevertheless, some studies, mainly devoted to the electrooxidation of acetaldehyde, can be cited. Effects of the nature of the catalytic electrode or the length of the carbon chain were considered in some studies.

The oxidation of acetaldehyde or other aliphatic aldehydes on platinum electrodes occurs at relatively high potentials compared to those at which the oxidation of alcohols occurs. There is no oxidation in the double-layer region, while two oxidation peaks generally appear in the oxygen region. This simple observation means that the mechanism of the electrooxidation of aldehydes should be different from that of alcohols. In the latter case, oxidation of the electrode surface, leading to the formation of an adsorbed oxygen layer, inhibits the electrooxidation process, whereas such adsorbed oxygenated species seem to be necessary in the case of aldehyde oxidation. The main products of electrooxidation of aldehydes are the corresponding carboxylic acids.

On other metals, the electrooxidation of aldehydes also forms mainly carboxylic acids, with by-products resulting from the Cannizzaro reaction. On metals such as nickel or copper, oxide layers are necessary for the electrooxidation to proceed but on gold or silver, the oxidation takes place on the oxide-free surface.

Adsorption of acetaldehyde leads to a partial dissociation of the molecule, as proved recently by spectroscopic measurements. Production of carbon dioxide, detected by DEMS and ECTDM, and the presence of linearly adsorbed CO, detected by EMIRS, both proved without doubt the breaking of the C—C bond. Adsorbed CO again plays the role of a catalytic poison. These observations are corroborated by experiments on platinum single crystals. However, when the concentration of acetaldehyde is not too low, the main product formed remains acetic acid.

The electrooxidation of acetaldehyde is greatly enhanced by the presence of foreign adatoms on platinum. Oxygen-adsorbing adatoms (Ru, Ge, Sri, AS, and Sb) strikingly increase the current densities during the electrooxidation of aliphatic aldehydes (formaldehyde, acetaidehyde, and propionaldehyde). Non-oxygen-adsorbing adatoms (Bi mainly) also have significant effects. In the former case, the enhancement of platinum activity can be explained mainly by the bifunctional theory of electrocatalysis, but in the latter case inhibition of the formation of poisoning species can certainly explain the observed increase in current.

(iii) Reactivity of the Ketone Function

A ketone function usually represents the highest possible oxidation level, and in a compound where a ketone is the only functional group present, no mild electrooxidation can occur, So that degradation of the molecule will take place, associated with the breaking of C—C bonds. However, ketones are very stable in aqueous medium, and electrochemical degradation requires high positive potentials as well as long adsorption times. For instance, the oxidation of adsorbed acetone is only possible after a rather long adsorption time, around 30 minutes. Under these conditions, CO_2 was detected by DEMS during the positive sweep, in the oxygen region of the platinum electrode.

Oxidation of products containing ketone groups can occur, as will be discussed in Section III.3, which deals with multifunctional compounds.

Electroreduction of ketones is possible, but this topic is not directly relevant to the subject of this chapter.

(iv) Reactivity of the Carboxylic Acid Function

The adsorption of carboxylic acids on platinum is a reversible phenomenon, and significant dissociation of the acid molecule does not occur. This fundamental observation means that the electrooxidation of carboxylic acids is difficult in aqueous medium, even though the degradation of such molecules is possible. As a typical example, the oxidation of oxalic acid, either at platinum or at gold electrodes, does occur in acid medium, but only at high positive potentials, with breaking of the C—C bond, leading to the formation of carbon dioxide.

Reduction processes have been reported to occur in the hydrogen adsorption region at platinum electrodes.

Oxidation of Multifunctional Molecules

In the previous section, the reactivity of different functional groups was discussed, mainly from the point of view of their oxidation at catalytic electrodes. However, the environment of the functional group plays a very important role in its reactivity. The influence of the position of the reactive group in the carbon chain is clearly evident for the oxidation of aliphatic monoalcohols, such as the butanol isomers. The length of the carbon skeleton is also an important factor. Nevertheless, the presence of (at least) one other functional group in the molecule may modify greatly its electrooxidation. While there is a very large variety of such molecules, some typical examples will be given below, dealing with molecules having practical interest, particularly in fuel cells and in electroorganic synthesis.

(i) Oxidation of Diols and Polyols

Due to the good solubility of polyols in water, and the relatively good reactivity of alcohol groups, the number of studies concerning the electrooxidation of polyols in aqueous medium is rather large.

The first member of the series, ethylene glycol, is probably the most promising fuel among those polyols with more than one carbon atom, even though many problems still remain unsolved for its use in an electrochemical power source. The electrooxidation of ethylene glycol has been widely studied on platinum and gold and on adatom-modified electrodes. The influence of the electrode surface structure has also been observed with platinum single crystals.

Ethylene glycol is easily oxidized on platinum electrodes in acid or alkaline medium. The current densities in the latter case are rather high (reaching around 7 mA cm-2). With gold, in alkaline medium, the electrocatalytic activity observed is also very significant (~12 mA cm^{-2}). The poisoning phenomena previously observed in the oxidation of monoalcohols are also seen with ethylene glycol.

As an example, in acid medium the oxidation of ethylene glycol at platinum electrodes starts with a shoulder at around 0.5 V versus RHE and with a main peak at 0.8 V. However, if the lower limit of the potential is shifted positively, the shoulder becomes a real peak showing weak poisoning phenomena. The meaning of such an observation is clear: in the hydrogen region, a poisoning species is formed by dissociative adsorption of ethylene glycol, partially blocking the electrode active sites. Spectroscopic measurements by EMIRS showed the presence of adsorbed CO, responsible for the poisoning of the platinum surface. Thus, in acid medium, CO is mainly linearly bonded to the electrode surface, while in alkaline medium bridge-bonded CO is dominant.

The presence of adatoms on platinum surfaces leads to a very large enhancement effect on the electroreactivity of ethylene glycol, especially in alkaline medium. Oxidation current densities as high as 100 mA cm^{-2} were observed with lead or bismuth adatoms. In acid medium, the effect is much weaker, and this can be related to the adsorption on the catalytic surface. As seen for formic acid adsorption on rhodium, lead adatoms occupy two adsorption sites and consequently modify strongly the coverage of the electrode by bridge-bonded CO. It was concluded that bridge-bonded CO is probably the species responsible for electrode poisoning. The electrooxidation of ethylene glycol at platinum in alkaline medium probably involves the same phenomenon, bridge-bonded CO being the main poisoning species present on the surface in this case.

On gold electrodes, the effect of adatoms is very small, which confirms that poisoning phenomena are quite absent in this case.

The reaction products encountered in the electrooxiclation of ethylene glycol are mainly C2 compounds, oxalic acid being the last stage of oxidation without breaking of the C—C bond. However, some products resulting from C—C bond breaking (HCHO, HCOOH, and CO$_2$) have also been detected by liquid and gas chromatographies.

Ethylene glycol electrooxidation is very sensitive to the structure of the electrode surface. As in the case of the oxidation of monoalcohols, the Pt(110) orientation appears to be the most active plane, but also the most poisonable one.

Despite the problems involved, ethylene glycol represents a possible fuel for an electrochemical power source, and practical systems have been studied using plurimetallic anodes. On the other hand, selective oxidation of ethylene glycol to glycolaldehyde was recently obtained at platinum electrodes by carefully choosing the oxidation potential and by modifying the electrode activity by introduction of foreign metal adatoms, an interesting result for electroorganic synthesis.

Studies on the electrooxidation of other diols are much more scarce. The two isomers of propanediol have however been considered by several research groups. On platinum, the electrooxidation of the two propanediol isomers follows different routes. 1, 2-Propanediol behaves very similarly to ethylene glycol, but with smaller current densities, whereas 1, 3-propanediol has a general behavior similar to that of a monoalcohol (propanol or butanol).

Numerous reaction products are formed during the prolonged electrolysis of such compounds. For example, during the oxidation of 1 2-propanediol at a platinum electrode, two main products have been identified by HPLC: lactic acid ($CH_3CHOHCOOH$) and hydroxyacetone (CH_3COCH_2OH). In addition, pyruvic acid ($CH_3COCOOH$) was also found, but in a smaller quantity. Lighter products, such as acetic acid and even formic acid, formed as the result of the breaking of the C—C bond, were also detected.

The electrooxidation of other polyols has also been investigated, such as that of glycerol and sorbitol. Such molecules have been often considered as interesting raw materials for electrosynthesis. Some oxidation products of diols may be of practical interest. As a typical example, the transformation of 2, 3-butanediol to 2-butanone can be achieved by oxidation of butanediol to acetoin followed by reduction to 2-butanone.

(ii) Oxidation of Other Multifunctional Compounds

Multifunctional compounds are diverse and display a variety of molecular structures. Numerous studies have been devoted to the electrooxidation of molecules containing some combination of alcohol, aldehyde, acid, and ketone functional groups.

Among these molecules, oxalic acid glyoxal, glyoxylic acid, lactic and pyruvic acids, and mesoxalic acid should be mentioned. In fact, these compounds are of interest for several very different reasons.

Some of them, such a diacids, are only oxidizable by breaking of C-C bonds, leading generally to carbon dioxide. Typical examples are oxalic and mesoxalic acids, which are easily oxidized to CO_2 in acid medium on platinum, but only at relatively high potentials corresponding to the oxidation of the platinum surface.

In the case of other compounds, such as glyoxal and lactic acid, reaction products corresponding to selective oxidation are interesting for electrosynthesis purposes. To perform such nondestructive mild oxidation necessitates finding convenient experimental conditions (in terms of potential and electrode structure) that avoid breaking of C—C bonds.

(iii) oxidation of Monosaccharides

The electrooxidation of glucose has been extensively studied, because of interest in the use. of this reaction, in biosensors and implanted fuel cells for pacemakers. Glucose is a typical product of the biomass, and practical electrosyntheses, using glucose as a raw material, have been proposed, such as production of gluconic acid by electrooxidation or of sorbitol by electroreduction. The simultaneous production of both gluconic acid and sorbitol in the same undivided cell was also carried out in order to evaluate paired synthesis.

The main reaction product formed during the electrooxidation of glucose is gluconolactone, either at platinum electrodes in acid medium or at gold electrodes in alkaline medium. Gluconolactone is further transformed by hydrolysis into gluconic acid. The first step in the reaction is the dehydrogenation of the C_1 carbon, which occurs at rather low potentials at a platinum electrode. A recent study by in situ FTIR spectroscopy showed that the platinum electrode surface is covered, at low potentials, by linearly adsorbed carbon monoxide coming from the partial dissociation of the glucose molecule. This adsorbed CO species is probably responsible for the electrode poisoning.

A significant effect of the electrode structure was observed with platinum in acid medium, and also with gold in alkaline medium. The three low-index single-crystal planes of platinum exhibit higher activity than polycrystalline platinum, and the Pt(111) orientation appears to be less sensitive to poison formation over the whole pH range.

The influence of foreign metal adatoms on the electrooxidation of glucose at platinum has been studied. Adatoms such as Tl, Pb, and Bi enhance greatly the activity of platinum in acid, as well as in alkaline, medium. These effects are generally interpreted in terms of a decrease of electrode poisoning.

The other monosaccharides are much less studied, but seem also less reactive. As an example, fructose is only slightly oxidized at platinum electrodes in acid medium, but more significantly at gold electrodes in alkaline medium.

OUTLOOK FOR THE OXIDATION OF SMALL ORGANIC MOLECULES AT CATALYTIC ELECTRODES

If one considers the tremendous development of techniques relevant to interfacial electrochemistry and the efforts which have been made in the past ten years to track the adsorbates on electrodes, it would not be an exaggeration to say that some new concepts have been definitively established.

Thus, insofar as the oxidation of small aliphatic organic molecules is concerned, the following points can be put forward:

(i) Adsorbates belong to two groups:

• *reactive species*, which are characterized by a weak adsorption to the surface and a rather short lifetime. Kinetic intermediates 'are supposed to belong to this group, 6ut none has been detected yet;

• *poisoning species*, which are characterized by a strong adsorption to the electrode surface and, therefore, high stability. CO$_{ads}$ is the most common poison. It can be formed by chemisorption of all small organic molecules in acid medium, but only by some of them in alkaline medium.

(ii) Different adsorbates are present simultaneously at the electrode surface. Furthermore, the active surface, that is, the catalytic surface under reaction, is not static. It should be recognized that the populations of adsorbates vary with time even if the experimental parameters remain fixed. Thus, poisoning phenomena are explained in terms of competitive adsorption. The more strongly adsorbed species compete and displace the weak adsorbates, until total deactivation of the surface occurs.

(iii) A limited number of adsorbates are implicated in the chemisorption of small organics. For, instance, Fig. 34 helps in understanding the correlation between the adsorbates formed by chemisorption of C, compounds.

(iv) *Surface arrangements* have to be considered to take into account the crystallography of active sites and the local number of adsorption sites necessary per adsorbate, as a result of which *electrocatalytic reactions are surface sensitive*. Except when the size of adsorbates increases to the point that steric effects become dominant, the behavior of a polycrystalline surface can be modeled by the weighted contributions of its basal planes. However, at the surface, further rearrangements and disproport reaction may occur as well, at least under some conditions.

(v) There is as yet *no general theory to explain the role of adatoms*. In some experiments, adatoms act by bringing together the oxygenated species necessary to trigger the electrochemical process. In other experiments, they simply act by selectively displacing the poisons or by modifying the way in which the adsorbates are bonded to the surface. Steric effects can also be invoked.

(vi) Developed surfaces and catalysts dispersed in conducting matrices are markedly less poisoned than smooth surfaces. They offer interesting possibilities for practical applications.

(vii) In the case of *multifunctional molecules*, it can be roughly said that each functional group has its own reactivity at a given surface in a given medium. Additive behavior may be expected, provided that the geometries of the adsorbates and of the surface allow the simultaneous reactions to occur.

This short summary of progress in the understanding of electrocatalytical reactions also shows the direction for future research. In particular, more work is needed on controlled model surfaces, as they are the basis of any theory.

Experimentally, now that sufficient sensitivity has been achieved for adsorbate detection, there is no doubt that further progress will come from advanced techniques relying on detection with much shorter time constants. Such techniques are required in order to study the formation of kinetic intermediates, which is the key to elucidating mechanisms.

Similarly, now that the reactive and poisoning species of catalytic surfaces have begun to be identified, further progress should be made in the development of highly stable practical catalytic electrodes. Applications will concern their industrial use, not only for electrochemical power sources burning organic fuels directly, but also for electrosynthesis processes.